T0327570

VALIDATING
CHROMATOGRAPHIC
METHODS

VALIDATING CHROMATOGRAPHIC METHODS
A Practical Guide

DAVID M. BLIESNER

WILEY-
INTERSCIENCE

A JOHN WILEY & SONS, INC., PUBLICATION

Library of Congress Cataloging-in-Publication Data:

Bliesner, David M.
    Validating chromatographic methods : a practical guide/by David
M. Bliesner.
        p. cm.
    Includes bibliographical reference and index.
    ISBN-13: 978-0-471-74147-3
    ISBN-10: 0-471-74147-7 (acid-free paper)
    1. Chromatographic analysis—Validity. 2. Science — Methodology.
I. Title.
QD79. C4B57 2006
543′ 8 — dc22   2005036658

CONTENTS

PREFACE

Delphi Analytical Services, Inc. has spent the past several years helping companies in the pharmaceutical industry improve their level of compliance with current good manufacturing practices (CGMPs). This involvement has included large and small companies who have already been subject to regulatory action from the Food and Drug Administration (FDA) as well as companies who are taking preventative measures to avoid regulatory action. As part of this effort, a significant amount of time has been spent reviewing analytical and bioanalytical methods and methods validation documentation.

Unfortunately, our experience leads us to conclude that despite improved guidance from the FDA, leadership by the International Conference on Harmonization (ICH), and the plethora of validation experts and courses, a substantial need still exists for education, training, and periodic retraining in the field of methods validation.

Make no mistake, analytical methods validation is not a trivial undertaking. And like snowflakes, no two are exactly alike. However, our experience has brought us in contact with what we consider to be the best practices in the industry. In addition, we have seen some of the mistakes that often degrade the overall quality of the finished product: A validated, transferable, analytical method that will serve its end users for an extended period of time with minimal complications.

In our experience, very few labs have it "all together" and execute *all* the components of a methods validation well. This guide was written with the intent to bring order to the potentially chaotic process of methods validation.

If you are new to methods validation, we hope to provide you with enough practical information and tools to keep you from having to reinvent the wheel by having to develop your own systems and to attack methods validation from scratch. If you are experienced with methods validation, we hope you will use this guide to upgrade and improve your existing systems.

This guide focuses on chromatographic methods validation, specifically high performance liquid chromatographic (HPLC) methods validation. This approach was chosen in that HPLC is by far the most common analytical technique used in modern pharmaceutical analytical R&D/QC laboratories. However, the concepts are in many cases directly applicable to validation of other analytical techniques as well.

In that CGMPs are always changing (hence the "C" meaning "current") and the industry is always improving and upgrading its best practices, we encourage your feedback and comments so that we can keep this guide in line with the best practices of the industry. We look forward to your input and hope this guide assists you in your continuing quest for quality.

DAVID M. BLIESNER

CHAPTER 1

OVERVIEW OF METHODS VALIDATION

1.1 WHAT IS METHODS VALIDATION?

The FDA in its most recent publication, *Guidance for Industry on Analytical Procedures and Methods Validation*, states:

> **Methods validation is the process of demonstrating that analytical procedures are suitable for their intended use. The methods validation process for analytical procedures begins with the planned and systematic collection by the applicant of the validation data to support analytical procedures. [1]**

What the FDA does not say is that the actual validation component of the methods validation process should be the culmination of a well-organized, well-planned, and systematically executed process that includes method development, prevalidation studies, and finally, methods validation itself. Gone are the days where one did methods development/validation concurrently. Validation is the end game where few surprises and deviations are expected. Validation is executed with a formal, approved and signed methods validation protocol in place which has been reviewed by the quality assurance (QA) unit. Validation is complete when you:

(1) Demonstrate that you have met all the acceptance criteria.
(2) Clearly document the results in a CGMP compliant fashion.

Validating Chromatographic Methods. By David M. Bliesner
Copyright © 2006 John Wiley & Sons, Inc.

(3) Show how you met the acceptance criteria in a final methods validation report, including references to raw data, all of which have been reviewed and approved by the appropriate personnel including peers, management, and QA.

Some would even argue that the validation process is not complete until the methods are successfully transferred to their end-user laboratories.

This sounds like a daunting task. And to be completely honest—it is. There is nothing trivial or easy about methods validation. It takes time, resources, and rarely goes as easy as you think it's going to go. Methods validation is part science, part art, and a lot of bookkeeping and accounting. To be brutally honest, too few laboratories do a very good job executing all the components.

Due to the magnitude of the task, the time, and the perceived costs, many laboratories try to cut corners. At a minimum, this results in deviations from the protocol which no longer can be "arm waved" away. The FDA expects you to scientifically address failures as you would any other laboratory investigation. This takes more time and effort and often results in delays in the validation timeline. In the worst case, you end up validating a method that is transferred to quality control (QC) labs worldwide, and ends up being the root cause of untold laboratory investigations. It is hypothesized that many of the problems discovered during root-cause analysis of out-of-specification results (OOS) are a direct result of poorly or partially validated methods.

The sections that follow provide a road map and the tools to guide and assist you to properly and efficiently validate your chromatographic methods, ensuring your validated methods do not become the root cause of your future laboratory investigations.

1.2   STEPS IN THE CHROMATOGRAPHIC METHODS VALIDATION PROCESS

The process of validating chromatographic methods can be broken down into four steps. These steps include:

- 1. Method evaluation and further method development,
- 2. Final method development and trial methods validation,
- 3. Formal methods validation, and
- 4. Formal data review and report issuance.

Figure 1.1 graphically represents the process.

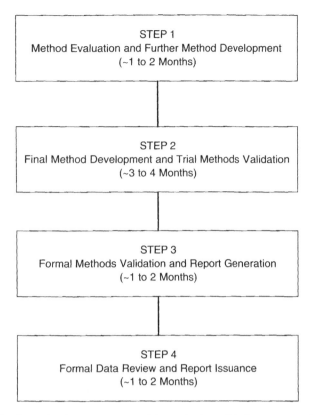

FIGURE 1.1   Steps in the chromatographic methods validation process. Total time for method evaluation, validation, data/documentation review, and reporting is approximately 6 to 10 months.

The estimates given in Figure 1.1 are only that—estimates. So many variables exist during methods validation that it makes it very difficult, if not impossible, to give an accurate prediction of the length of the validation project. There is an enduring myth in the pharmaceutical industry that it should *only take six weeks* to validate a method. The formal validation portion itself should only take about six weeks, but the preparation and documentation take significantly more time. Although it can be done, a complete validation, including proper data review and documentation, even for an established product with known properties, has not been satisfactorily completed in such a short time. Industry professionals believe that this is why there are so many bad methods in use today.

Details of the design and implementation for each step are described in the remaining sections of this guide. In addition, many of the tools, templates, and examples needed to complete methods validation are included in the appendices. As many examples, based on real-world scenarios, as practical have been provided to give you a framework to validate your own methods. In particular, the following key elements which constitute a methods validation program have been provided:

- A template for a methods validation standard operation procedure (SOP) example
- A template for a standard test method example
- A template for a methods validation protocol example
- A template for a methods validation report example

Each template for these examples represents a significant body of knowledge and experience. It would require a substantial amount of your time to create these templates independently. Modify them and use them to best fit the practices within your organization.

1.3 HOW TO USE THIS GUIDE

Before proceeding, here are some suggestions on how to use this guide. Although this document is a practical guide, it is by no means a technical cookbook on how to validate an analytical method. This would be impossible since every method has its own unique idiosyncrasies. Therefore, it is recommended that you take the following approach to best apply this guide:

- Familiarize yourself with the standard methods validation terms listed in the glossary.
- Read and understand the guide section titled "Components of a methods validation."
- Review the template for the methods validation SOP example.
- Review the template for the test method example.
- Review the template for the methods validation protocol example.
- Read the template for the methods validation report example.
- Read and understand the flowcharts and checklist related to methods validation in steps 1– 4.
- Develop your own systems and templates by adapting the systems and templates presented in this guide to your laboratory as appropriate.
- Train your chemists on the systems.
- Implement your systems.

1.4   ADDITIONAL POINTS TO CONSIDER WHEN VALIDATING CHROMATOGRAPHIC METHODS

In addition to the nuts and bolts of methods validation, many soft skills exist that will improve your chances of success during validation. Therefore, as you apply this guide to methods validation in your own laboratory, please keep the following points in mind.

Do Not Underestimate the Value of Planning and Organization

Much of methods validation is bookkeeping—both figuratively and literally. Therefore, much of the success of methods validation is dependent upon the amount of effort and attention to detail made in steps 1 and 2. The framework of what constitutes a methods validation is predetermined by FDA, ICH, and current industry practice. Think systematically and work with the end in mind. Therefore, by prioritizing and planning your work carefully, allocating your resources efficiently, coupled with good supervision and communications, you will significantly enhance your chances of a positive and timely outcome.

Make It Simple, Keep It Simple and Remember Your End Users

Chromatographic methods are often developed and validated by analytical research-and-development scientists who are not the end users of the method. Because of this, end user requirements are often not taken into consideration, which may lead to an overly complex and scientifically elegant method that will be "thrown over the wall" to the quality control (QC) chemists during technology transfer. Despite the myth, QC chemists and technicians are usually technically sound and well educated. However, they often work in a pressure cooker environment where the complexity and nuances of a method will only make their life more difficult. They need to get product out the door, with minimal complications and effort. Therefore, methods should be made as simple and robust as possible, with the end user's needs in mind.

QA *Is* Your Friend

It was once said the best quality assurance (QA) person is the one who eats his or her lunch alone because none of the chemists wants to talk with him or her. Unfortunately there is often an unhealthy tension that develops between QA and the lab. You should make every effort to reduce such tensions during the methods validation process. Remember, in the future your work will be reviewed cold, without coaching from you, by an FDA reviewer. Therefore, you need to develop and present a complete and accurate account of your work which raises few questions. So look at QA as your first line of defense. If you can't make them understand your work, how will the FDA understand it?

Don't Underestimate the Value of Experience

As stated, no two methods validations are the same, but each is very similar. This means that someone somewhere has probably encountered the same problems you are encountering. As part of your step 1 planning or troubleshooting, consider who might have experience with your work or work that is similar to what you are doing. Look within your own organization at previous validations. Look to vendors outside your organization such as the API manufacturer, your reagent supplier, and your chromatographic equipment and material supplier. Go to the library, search the literature, and tap into your network within the industry. Chances are someone has an answer or even *the answer* to your question. Avoid the "not invented here syndrome." Don't fall in love with your own work and skills to the exclusion of other good ideas. Ph.D. level R&D scientists are particularly bad about this. Remember the goal and resist the temptation of creating another dissertation research project.

Common Sense Is an Uncommon Virtue

During the course of the validation process perform what we call periodic sanity checks. Stop and ask yourself: Do these results make sense? Does this solution to the problem make sense for my end users? Am I headed in the right direction? Don't be shy about talking to your end users as well.

Mistakes Are Made Under Pressure

The validation of bad methods invariably comes by having to perform the validation under pressure. Again, this is why it's so important to expend a significant effort on steps 1 and 2 of the validation process. When you get to month six and still don't have a functional (let alone validated) method, pressure will make you get the methods out to your end users when they never should have made it out of your lab. People make mistakes, but people make more significant mistakes under pressure.

Realize the Impact of Your Successes and Your Failures

The lifetime of a validated chromatographic method can span decades. The method may be used in laboratories all over the world. The financial and resource usage impact of the method can be substantial. Keep this in mind as you validate your methods. The rule is once the method is validated and transferred, the chances of changing anything significantly with the method are very limited. From a regulatory standpoint, this situation raises a significant number of questions. From a practical standpoint, it costs too much and takes away resources from the next project. Be mindful of these points and do it right the first time. Remember:

Validating an analytical method may be the most important task you will perform during your tenure with your company.

CHAPTER 2

COMPONENTS OF METHODS VALIDATION

2.1 BACKGROUND

Chromatographic methods validation is subdivided into four categories, generally recognized via the United States Pharmacopoeia (USP). These categories include:

Category   I. Validation of analytical methods for assay
Category II. Validation of analytical methods for impurities and degradants
Category III. Validation of analytical methods for dissolution
Category IV. Validation of analytical methods for identification

Although clearly separated, validation can encompass more than one category simultaneously, depending on the workplan or resources available. For example, a method may be (and frequently is) validated concurrently for assay and for impurities and degradants. This is true because many of the validation characteristics tested during assay validations are the same as for impurities and degradants. A description of the validation characteristics are listed in the following sections.

Validating Chromatographic Methods. By David M. Bliesner
Copyright © 2006 John Wiley & Sons, Inc.

2.2   APPROACH

The following are typical analytical performance characteristics which may be tested during methods validation:

- *Accuracy*
- *Precision*
  - Repeatability
  - Intermediate precision
- *Specificity*
- *Detection limit*
- *Quantitation limit*
- *Linearity*
- *Range*
- *Robustness*
- *System suitability determination*
- *Forced degradation studies*
- *Solution stability studies*
- *Filter retention studies*
- *Extraction efficiency studies*
- *Additional methods validation information*
  - Representative instrument output
  - Representative calculations
  - Listing and characterization of known impurities
  - Degradation pathways (if known)

A more detailed definition of each characteristic is given in the following subsections.

2.2.1   Accuracy

Accuracy is the nearness of a measured value to the true or accepted value. It provides an indication of any systematic error or bias in the method. For an unbiased method, a theoretical plot of measured value versus true value can be described by a mathematical function. In the pharmaceutical industry, this is typically a straight line with a given slope and zero intercept. It follows that the accuracy of a biased method varies with the analyte concentration according to the types of systematic errors.

During the validation, accuracy is determined by measuring the recovery of the active component from a drug product matrix or by directly measuring the active pharmaceutical ingredient (API). Typically studies involve spiking the drug product placebo matrix with API in amounts equal to the nominal finished dosage strength. This spiking is either by adding of standard solutions or dry spiking API into the matrix followed by complete mixing.

2.2.2 Precision

Precision consists of two components: repeatability and intermediate precision. Repeatability is the variation experienced by a single analyst on a single instrument. Repeatability does not distinguish between variation from the instrument or system alone and from the sample preparation process. During the validation, repeatability is performed by analyzing multiple replicates of an assay composite sample by using the analytical method. The recovery value is calculated and reported for each value.

Intermediate precision refers to variations within a laboratory such as different days, with different instruments, and by different analysts. This was formerly known as ruggedness. During the validation, a second analyst repeats the repeatability analysis on a different day using different conditions and different instruments. Recovery values are calculated and reported. A statistical comparison is made to the first analyst's results.

2.2.3 Specificity

Specificity is the ability to assess unequivocally the analyte in the presence of components that may be expected to be present such as impurities, degradation products, and excipients. There must be inarguable data for a method to be specific. Specificity = measures only the desired component without interference from other species that might be present; separation is not necessarily required.

To determine specificity during the validation blanks, sample matrix (placebo), and known related impurities are analyzed to determine whether interferences occur. Specificity is also demonstrated during forced degradation studies.

The term "selectivity" is sometimes used interchangeably with specificity. Technically, however, there is a difference. Selectivity is defined as the ability of the method to separate the analyte from other components that may be present in the sample, including impurities. Selectivity is separate and shows every component in the sample. Therefore, one could have a method that is specific, yet it may not be selective. For instance, an ion selective electrode may be specific (e.g., is used to measure a single species in sample matrix), yet not be selective (e.g., doesn't separate and identify all components present).

2.2.4   Detection Limit

The detection limit (DL) or limit of detection (LOD) of an individual procedure is the lowest amount of analyte in a sample that can be detected but not necessarily quantitated as an exact value. The LOD is a parameter of limit tests (i.e., tests that only determine if the analyte concentration is above or below a specification limit).

In analytical procedures such as HPLC that exhibit baseline noise, the LOD can be based on a signal-to-noise (S/N) ratio (3:1), which is usually expressed as the concentration (e.g., percentage, parts per billion) of analyte in the sample. There are several ways in which it can be determined, but it usually involves injecting samples, which generate an S/N of 3:1, and estimating the DL.

2.2.5   Quantitation Limit

The quantitation limit (QL) or limit of quantitation (LOQ) of an individual analytical procedure is the lowest amount of analyte in a sample that can be quantitatively determined with suitable precision and accuracy. The quantitation limit is a parameter of quantitative assays for low concentrations of compounds in sample matrices and is used particularly for the determination of impurities and/or degradation products. It is usually expressed as the concentration (e.g., percentage, parts per million) of analyte in the sample.

For analytical procedures such as HPLC that exhibit baseline noise, the LOQ is generally estimated from a determination of S/N ratio (10:1) and is usually confirmed by injecting standards which give this S/N ratio and have an acceptable percent relative standard deviations (%RSDs) as well.

2.2.6   Linearity

Linearity evaluates the analytical procedure's ability (within a give range) to obtain a response that is directly proportional to the concentration (amount) of analyte in the sample. If the method is linear, the test results are directly or by well-defined mathematical transformation proportional to the concentration of analyte in samples within a given range. Linearity is usually expressed as the confidence limit around the slope of the regression line. The line is calculated according to an established mathematical relationship from the test response obtained by the analysis of samples with varying concentrations of analyte. Note that this is different from *range* (sometimes referred to as *linearity of method*), which is evaluated using samples and must encompass the specification range of the component assayed in the drug product.

During validation, linearity may be established for all active substances, preservatives, and expected impurities. Evaluation is usually performed on standards.

2.2.7 Range

Range is defined as the interval between the upper and lower concentrations (amounts) of analyte in the sample (including these concentrations) for which it has been demonstrated that the analytical procedure has a suitable level of precision, accuracy, and linearity.

Range is normally expressed in the same units as test results (e.g., percent, parts per million) obtained by the analytical method.

During validation, range (sometimes referred to as *linearity of method*) is evaluated using samples (usually spiked placebos) and must encompass the specification range of the component assayed in the drug product.

2.2.8 Robustness

Robustness is defined as the measure of the ability of an analytical method to remain unaffected by small but deliberate variations in method parameters (e.g., pH, mobile-phase composition, temperature, and instrument settings) and provides an indication of its reliability during normal usage. This is an important parameter with respect to the transferability of the method following validation.

Determining robustness is a systematic process of varying a parameter and measuring the effect on the method by monitoring system suitability and/or the analysis of samples. It is part of the formal methods validation process.

2.2.9 System Suitability Determination

System suitability is the evaluation of the components of an analytical system to show that the performance of a system meets the standards required by a method. A system suitability evaluation usually contains its own set of parameters. For chromatographic assays, these may include tailing factors, resolution, and precision of standard peak areas, and comparison to a confirmation standard, capacity factors, retention times, and theoretical plates.

During validation, where applicable, system suitability parameters are calculated, recorded, and trended throughout the course of the validation. Final values are then determined from this history.

2.2.10 Forced Degradation Studies

Forced degradation or stress studies are undertaken to deliberately degrade the sample (e.g., drug product, excipients, or API). These studies are used to evaluate an analytical method's ability to measure an active ingredient and its degradation products, without interference, by generating potential degradation products.

During validation, samples of drug product (spiked placebos) and drug substance are exposed to heat, light, acid, base, and oxidizing agent to produce approximately 10% to 30% degradation of the active substance. The degraded samples are then analyzed using the method to determine if there are interferences with the active or related compound peaks. Forced degradation studies can be time consuming and difficult because it is often difficult to generate the proper level of degradation. Also, a certain amount of logic needs to be applied to extrapolate the results of these studies to what might be seen during actual stability studies.

2.2.11 Solution Stability Studies

During validation the stability of standards and samples is established under normal benchtop conditions, normal storage conditions, and sometimes in the instrument (e.g., an HPLC autosampler) to determine if special storage conditions are necessary, for instance, refrigeration or protection from light. Stability is determined by comparing the response and impurity profile from aged standards or samples to the response and impurity profile of freshly prepared standards.

2.2.12 Filter Retention Studies

Filter retention studies are a comparison of filtered to unfiltered solutions during a methods validation to determine whether the filter being used retains any active compounds or contributes unknown compounds to the analysis. Blank, sample, and standard solutions are analyzed with and without filtration. Comparisons are made in recovery and appearance of chromatograms.

2.2.13 Extraction Efficiency Studies

Extraction efficiency is the measure of the effectiveness of extraction of the drug substance from the sample matrix. Studies are conducted during methods validation to determine that the sample preparation scheme is sufficient to ensure complete extraction without being unnecessarily excessive. Extraction efficiency is normally investigated by varying the shaking or sonication times (and/or temperature) as appropriate during sample preparation on manufactured (actual) drug product dosage forms.

2.2.14 Additional Methods Validation
Information Often Required

In addition to these analytical performance characteristics, the following information is usually obtained or presented in the final validation report:

- Representative instrument output
- Representative calculations
- Listing and characterization of known impurities
- Degradation pathways (if known)
- Determination of relative response factor

Analytical methods performance characteristics are only part of the vocabulary that constitutes the language of methods validation. If you are new to the methods validation process, please review Appendix I, "Glossary of Methods Validation Terms," before proceeding.

2.3 THE BEST PLACE TO START

You must consider the use of the method and what validation characteristics are crucial for you to determine the proper use of the method. For example, if you are validating a stability indicating method for related compounds, then the following order is probably best:

1. Selectivity
2. LOD/LOQ
3. Forced degradation studies

If you have problems with selectivity, then you are dead before you start. You need to know this at the beginning. If you cannot detect a related compounds at the lowest level needed, then you won't be able to ensure you are meeting your specifications.

Perform a sanity check at the beginning so you don't have an *"oops"* halfway through your validation.

CHAPTER 3

STEP 1: METHOD EVALUATION AND FURTHER DEVELOPMENT

3.1 BACKGROUND

Although it has been stated that the FDA expects methods validation to be an independent and separate process from method development, it is sometimes very difficult for a laboratory to distinguish between the two functions. The purpose of step 1: Method evaluation and further method development is to assist in organizing your thoughts and guiding you in making the transition from method development to methods validation.

In some circumstances this transition will be straightforward and simple. For example, if a significant amount of research and development has been performed and a development report generated, step 1 will be a sanity check to make sure all the proper effort has been completed prior to prevalidation and formal validation studies (e.g., limited evaluation and no additional development are needed). However, if you have just been handed an older existing method that has not been validated to existing standards, step 1 needs to be executed with greater care and attention to detail (e.g., significant evaluation and perhaps substantially additional development are needed).

Regardless, we assume that by the time you are ready to begin step 1 you have performed a certain amount of method development yourself or have

Validating Chromatographic Methods. By David M. Bliesner
Copyright © 2006 John Wiley & Sons, Inc.

found a separation in the literature or from another source which fits your needs.

Step 1 does have a clear deliverable objective: To identify a method suitable for the end user's intended needs and stands a reasonable chance of being validated without complications. These criteria must be supported by data. Remember, the more initial effort, the better the entire validation will proceed and the higher the quality of the finished product.

It should be noted that in this guide, we chose not to spend significant time on the concept (and art) of method development. This is an entire topic unto itself and more of a scientific endeavor than an act of regulatory compliance. We simply offer a means to gather and organize information which can be used to develop or evaluate your method and make the transition from development to validation.

3.2 APPROACH

The basic elements of step 1 are:

- Methods validation commissioned in writing
- Methods validation team leader assigned
- Background information and end user requirements sourced, collected, and compiled
- Validation team leader selects validation team members
- Background information distributed for review and evaluation by the validation team
- Planning meetings held; work breakdown structure and project plan created
- Team member training conducted
- Project plan executed
- Development review report/summary report generated

Regardless of how you proceed, however, much of your validation efforts should be guided by your company's methods validation standard operating procedure (SOP). A template of a methods validation SOP example is presented in Appendix II and should be reviewed at this point if you are new to the methods validation process. Steps in this process are shown in Figure 3.1 and described in Table 3.1.

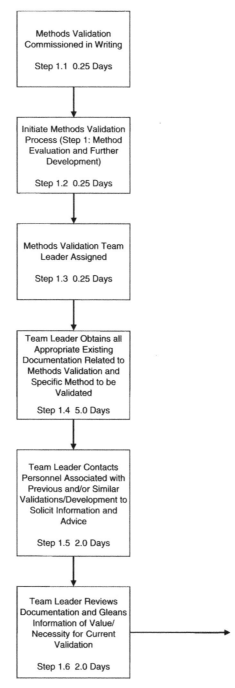

Components of Step 1:
Method Evaluation and Further Method Development
~1 to 2 Months

Methods Validation
Commissioned in Writing

Step 1.1 0.25 Days

Initiate Methods Validation
Process (Step 1: Method
Evaluation and Further
Development)

Step 1.2 0.25 Days

Methods Validation Team
Leader Assigned

Step 1.3 0.25 Days

Team Leader Obtains all
Appropriate Existing
Documentation Related to
Methods Validation and
Specific Method to be
Validated

Step 1.4 5.0 Days

Team Leader Contacts
Personnel Associated with
Previous and/or Similar
Validations/Development to
Solicit Information and
Advice

Step 1.5 2.0 Days

Team Leader Reviews
Documentation and Gleans
Information of Value/
Necessity for Current
Validation

Step 1.6 2.0 Days

FIGURE 3.1   Diagram of workflow for step 1: method evaluation and further method development.

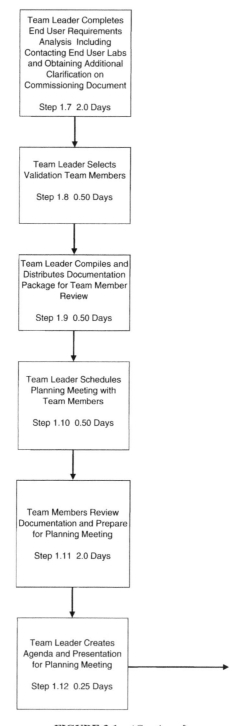

FIGURE 3.1 *(Continued)*

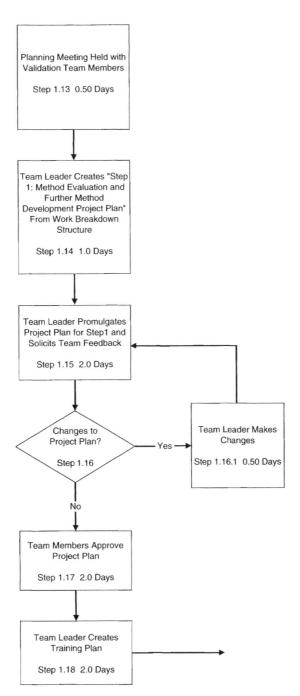

FIGURE 3.1   (*Continued*)

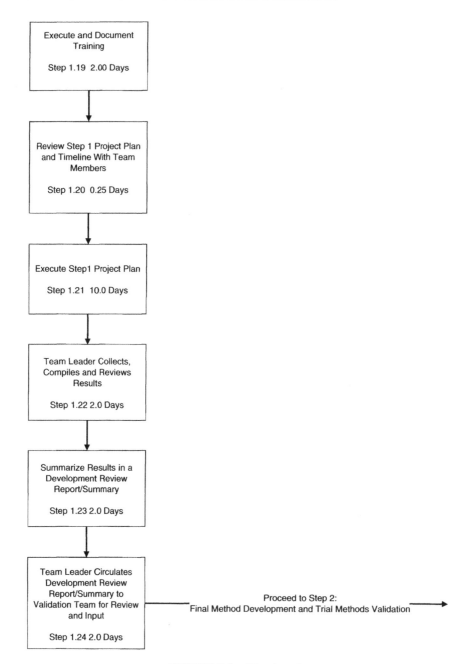

FIGURE 3.1   (*Continued*)

TABLE 3.1   **Explanation of Workflow Diagram Steps for Step 1:
Method Evaluation and Further Method Development**

Step	Description	Estimated Duration	Explanation
1.1	Methods validation commissioned in writing	0.25 day	The success of any project, including method validation, depends on management commitment and involvement. Therefore, it is important that management, within the organization responsible for methods validation, formally commission the validation in writing. As part of this commissioning, management should: (1) assign a team leader, (2) define the product type which will require the analytical methodology, (3) define the USP method category under which the method will be validated, (4) provide required start and end dates, (5) define method transfer goals and dates of transfer, and (6) determine whether the validation will be full, partial, or compendial.
1.2	Initiate methods validation process (step 1: method evaluation and further method development)	0.25 day	This should correspond to the start date determined by management and cited in the commissioning document.
1.3	Methods validation team leader assigned	0.25 day	As described in step 1.1, assignment of the team leader should be part of the methods validation commissioning procedure. The selection of a methods validation team leader is important. The team leader will be held accountable for the successful completion of the validation within the expected timeframe. An individual with the appropriate technical skills as well as project management and organizational skills is required. A good understanding of laboratory CGMPs and CGMP documentation requirements is also important.
1.4	Team leader obtains all appropriate existing documentation related to methods validation and specific method to be validated	5.0 days	The following documentation should be sourced if available and/or appropriate: (1) method development SOP, (2) method validation SOP, (3) method development reports, (4) API manufacturer technical reports and documents, (5) existing written methods, (6) USP, BP, and EP methods,

(Continued)

TABLE 3.1   *(Continued)*

Step	Description	Estimated Duration	Explanation
			(7) existing methods validation packages for this product or chromatographic separation, (8) existing methods validation packages for similar products or molecules, (9) ICH guidelines and FDA guidance documents, (10) selected text references such as *Analytical Profiles of Drug Substances and Excipients,* [3], (11) scientific literature search (paper or electronic) results for the product, drug substance, or related compounds, (12) queries of HPLC column vendors, HPLC supply vendors, etc., (13) other sources of information which may have physical-chemical properties of active and excipients including solubility, pKa, spectral properties, etc. (*Note:* See Appendix II for an example template of a methods validation SOP.)
1.5	Team leader contacts personnel associated with previous and/or similar validations/ development to solicit information and advice	2.0 days	As indicated in step 1.4, it is important to investigate if methods have been developed and/or validated previously for your molecules and products. Often work has already taken place within your company or another company. Do not pass up the opportunity to talk with the people who have already performed the work. In many cases, even if they are not from your company, they will be willing to share their experiences. In turn, they may be able to save you a lot of time.
1.6	Team leader reviews information of value/necessity for current validation	2.0 days	Once the information and feedback is obtained, the team leader gleans the pertinent information and compiles a concise documentation package for later distribution to the validation team.
1.7	Team leader completes end-user requirements analysis including contacting end-user labs and obtaining additional clarification on commissioning document	2.0 days	As part of the preparation process for methods validation, it is imperative to obtain feedback from the individuals who will be the end users of the method. Some of the questions which should be asked include: (1) Who will use the method (e.g., QC lab for routine testing, in-process lab)? (2) What types of equipment are available and are there any equipment limitations? (3) Where

TABLE 3.1 (*Continued*)

Step	Description	Estimated Duration	Explanation
			will the method be used (e.g., geographic locations and actual facility location)? (4) What supply restrictions may exist? (5) What expertise (e.g., the education, training, and experience levels of the end users) restrictions may exist? (6) What language barriers may exist? (7) Under which USP/ICH methods category does the method fall? (8) What validation characteristics are required? (Note: See Appendix III for an example of a template of an end-user questionnaire.)
1.8	Team leader selects validation team members	0.5 day	Some selection criteria for team members should include the following: (1) Is there a skill match to similar product/molecules and techniques? (2) Is there an experience match to similar products/molecules and techniques? (3) Are the personnel currently available to work on the project and can they handle the additional workload? (3) What is the employment status and person situation of the potential team members (e.g., they may be in line for a promotion which would take them away from the project)? (4) Do they have the ability to work closely with the other team members? The team member selection should be reviewed, approved, and then communicated through management. This will ensure conflicts no with personnel, scheduling, or research allocation.
1.9	Team leader compiles and distributes documentation package for team member review	0.5 day	This documentation package should include: (1) the commissioning document, (2) the current revision of the methods validation SOP, (3) existing development reports and documentation, (4) existing methods, and (5) the end-user requirements assessment.
1.10	Team leader schedules planning meeting with team members	0.5 day	The team leader should make an attempt to dovetail this meeting with the team members' existing schedules.

(Continued)

TABLE 3.1 *(Continued)*

Step	Description	Estimated Duration	Explanation
1.11	Team members review documentation and prepare for planning meeting	2.0 days	Team members should perform a thorough review of the package prior to attending the meeting. They should then generate a list of questions to present at the first team meeting.
1.12	Team leader creates agenda and presentation for planning meeting	0.25 day	The agenda should include: (1) team member introductions, (2) team member feedback on the end-user requirement responses, (3) timeline analysis, (4) quality of existing data and documentation, (5) team member experience with the product, molecules, or project, (6) team member training requirements, (7) team member work preferences (e.g., preferences for executing certain portions of the validation or use of specific analytical techniques), (8) a group brainstorming session to identify any potential difficulties prior to creating a project plan.
1.13	Planning meeting held with validation team members: Work breakdown structure is created	0.5 day	The purpose of the meeting is threefold: (1) to formally establish the team, (2) to inform team members on the details of the project, and (3) to have the team leader obtain feedback from the team members so the team leader can develop an accurate and realistic project plan and make appropriate work assignments.
1.14	Team leader creates step 1: method evaluation and further method development project plan from work breakdown structure	1.0 day	The project plan should be a straightforward document, such as a calendar, which delineates the tasks required to evaluate the method and/or the tasks required to perform additional methods validation. Assignment of team member responsibilities and start and stop dates are also included in the project plan. Although each project plan will be unique, some example tasks which may be included in the project plan include: An example "best case" scenario (e.g., method previously developed, complete method development report and package exists): (1) Evaluate development and documentation for API standard, related compound standards, placebo, and spiked

TABLE 3.1   (*Continued*)

Step	Description	Estimated Duration	Explanation
			placebo with existing method to confirm that separation is valid.
			(2) Evaluate chromatography presented in development documentation for upper and low ends of desired linear range.
			(3) Perform a sanity check on the separation with respect to compound solubility, stability of solutions, pK_a versus pH of the buffer, toxicity of materials, appropriateness of lambda max, etc.
			(4) Review end-user's requirements and match to method characteristics to ensure method will work for intended use.
			(5) Perform confirmatory separation with method and calculate fundamental chromatographic figures of merit, including LOD/LOQ if appropriate for API standard, related compounds standards, placebo, and spiked placebo as appropriate.
			An example "worst case" scenario (e.g., literature references for methods only):
			(1) Perform a sanity check of literature separation with respect to compound solubility, stability of solutions, pK_a versus buffer pH, toxicity of materials, appropriateness of lambda max, etc.
			Note: See Appendix IV for an example of a template of a method review checklist for summary sanity check question to be considered.
			(2) Perform a limited robustness study to include mobile phase adjustments, pH adjustments, temperature changes, change of column manufacturer, etc.
			(3) Perform LOD and LOQ tests as appropriate.
			(4) Review end-user requirements and match to method characteristics to ensure method will work for intended use.
			(5) Determine "go" or "no go" on the method as written.

(Continued)

TABLE 3.1   *(Continued)*

Step	Description	Estimated Duration	Explanation
			Note: At this stage, the team may have a substantial amount of information about the separation, or perhaps very little information. It depends entirely upon how much development work has been performed prior to this point.
			Note: This project plan is specifically designed for step 1 only.
1.15	Team leader promulgates project plan for step 1 and solicits team feedback	2.0 days	This gives the team members the opportunity to make corrections or modifications to the project plan for this phase of the validation.
1.16	Changes to project plan? If yes, team leader makes changes and forwards back to team members for feedback	0.5 day	If team members have made suggestions for changes, the team leader will make the changes as appropriate and recirculate the plan for additional feedback. Management should also be given the opportunity to review the draft project plan.
1.17	Team members approve project plan	2.0 days	Once all the changes have been made, the team should come to an agreement on the plan's content and move forward toward implementation. This acceptance should be shared with all team members and management.
1.18	Team leader creates training plan	2.0 days	Once the project plan is accepted, training of the team members needs to be executed. Some areas which may require training include: (1) SOP training/retraining, (2) analytical technique training/retraining, (3) training on methods validation protocol execution, (4) data capture and review procedures, (5) workflow procedure review, (6) data reporting procedure review, and (7) laboratory investigation reporting (LIR) procedures required during validation.
			Although these topics may not be applicable to the method evaluation and further development phase, they will be applicable during formal methods validation. Additional topics may be included during this training.

TABLE 3.1   (*Continued*)

Step	Description	Estimated Duration	Explanation
1.19	Execute and document training	2.0 days	Self explanatory
1.20	Review step 1 project plan and timeline with team members	0.25 day	Following training, and prior to executing the project plan, the plan and timeline should be reviewed with all team members one last time.
1.21	Execute step 1 project plan	10.0 days	Start date should be formally communicated to management and all team members.
1.22	Team leader collects, compiles, and reviews results	2.0 days	As data are obtained or an analysis of existing data is conducted by the team members during methods evaluation, the team leader begins compiling and organizing the findings to be included into a development review/summary report.
1.23	Summarize results in a development review report/summary	2.0 days	At this point, the team leader will generate a development review and similar summary report.
1.24	Team leader circulates development review report/summary to validation team for review and input	2.0 days	The development review or summary report is circulated among the team members for review and input and in preparation for a group review session. Proceed to step 2: final method development and trial methods validation.

CHAPTER 4

STEP 2: FINAL METHOD DEVELOPMENT AND TRIAL METHODS VALIDATION

4.1 BACKGROUND

At the conclusion of the tasks delineated in step 1, many organizations launch immediately into formal methods validation. In many cases this is a risky approach. Although the data may support the fact that the method will work for its intended use, the components of methods validation make it a very rigorous process, fully testing the ability for the method to be used consistently in many different laboratory environments.

Recall that the FDA expects validation to be the end game, where few surprises and deviations are encountered. Moreover, methods validation must be initiated and executed by creation and implementation of a methods validation protocol. If deviations or modifications from the protocol occur during validation, documentation is required to explain the deviations and support them with data. You cannot simply explain away failures or modify the protocol in order to make the experiments successful.

In many cases, failure to meet protocol acceptance criteria often leads to additional methods development. This in turn requires modification of the existing protocol or even creation of a new protocol with subsequent revalidation. Obviously this do-over cycle can take a tremendous amount of additional time

Validating Chromatographic Methods. By David M. Bliesner
Copyright © 2006 John Wiley & Sons, Inc.

and effort. Therefore, it is in the laboratory's best interest to make sure the method works as intended before executing the methods validation protocol. This leads us to step 2: final method development and trial methods validation.

The purpose of step 2 is to use the data obtained in step 1, to make a logical determination that the method can be validated, and then to create a trial or practice methods validation protocol and implement a trial methods validation via the protocol.

Since the trial protocol is not an official document and the data collected will not be included in any type of regulatory submission, the laboratory is free to modify the experiments and alter acceptance criteria as necessary. Although this may seem like overkill, experience shows that invariably difficulties are encountered during the first pass of a validation, and because the data are now part of a regulatory submission, the time and effort it takes to implement the do-over cycle easily surpass the time and effort required to perform a trial validation.

The clear deliverables for step 2: final method development and trial methods validation include:

1. A final methods validation protocol with established, reasonable, and obtainable acceptance criteria
2. A method which you know you can validate

If you have some previous experience conducting method development and/or performing methods validation, this approach may seem like overkill. However, experience shows that difficulties are invariably encountered during the first pass of a validation. Because of this, the value of a trial methods validation cannot be overemphasized. Therefore, the best guidance we can give to you is:

Do not gamble! Perform a trial methods validation first.

Regardless of your decision, templates of an example methods validation protocol and a standard test method are included for your review in Appendices VI and V, respectively.

4.2   APPROACH

The basic elements of step 2 are

- Validation team discusses development reports or summaries to assess whether further development work is needed, and to initiate work as appropriate
- Create trial methods validation protocol

- Create methods validation project plan
- Implement trial methods validation via protocol and project plan
- Collect and review data
- Create trial methods validation report
- Compare results obtained during trial validation versus the trial protocol
- Determine if method is validatable; modify experiments and protocol as necessary
- Execute additional experiments as necessary
- Proceed to formal methods validation

Steps in this process are shown in Figure 4.1 and described in Table 4.1.

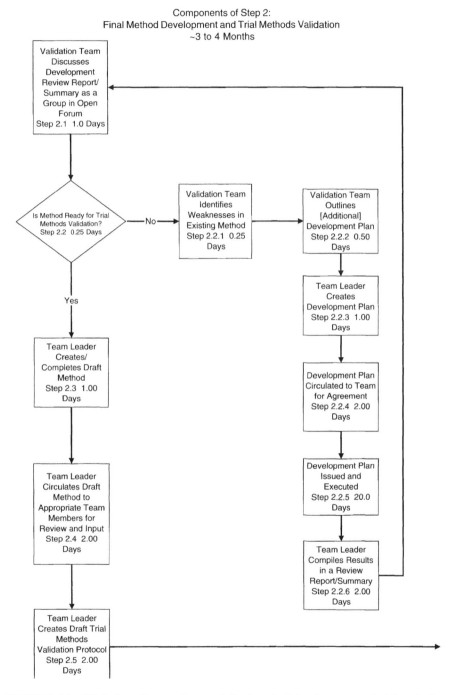

FIGURE 4.1 Work flow diagram for step 2: final method development and trial methods validation.

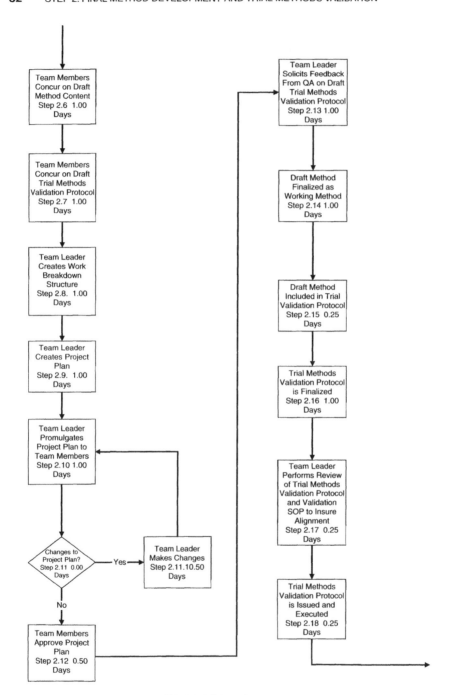

FIGURE 4.1   (*Continued*)

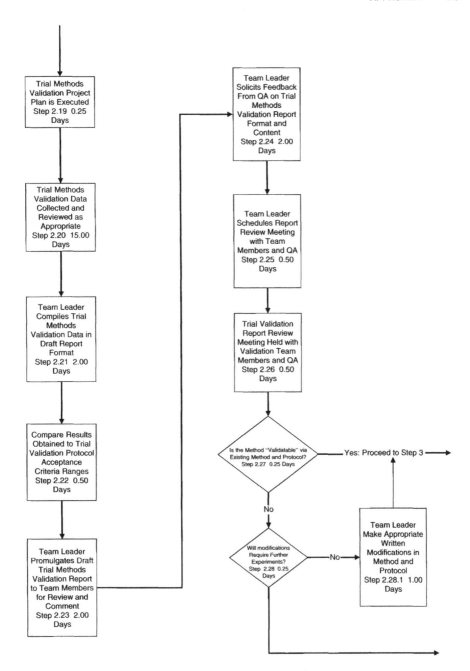

FIGURE 4.1   *(Continued)*

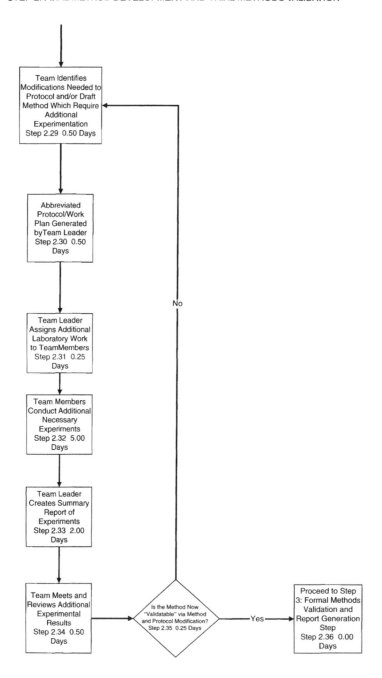

Total Estimated Time For Step 2 =75.0 Days
3.8 Working Months

FIGURE 4.1   (*Continued*)

TABLE 4.1   **Workflow Diagram for Step 2:**
Final Method Development and Trial Methods Validation

Step	Description	Estimated Duration	Explanation
2.1	Validation team discusses development review report/summary as a group in open forum	1.0 day	
2.2	Is method ready for trial methods validation?	0.25 day	Some criteria for a no-go decision may include: • Development work indicates that the method does not satisfy significant components of end-user requirements • Insufficient development data to support the decisions to validate the method • A significantly better method was discovered during step 1 based on literature search, vendor contacts, previous experience of team members, etc.
	If no, then		
2.2.1	Validation team identifies weaknesses in existing method	0.25 day	
2.2.2	Validation team outlines [additional] development plan	0.5 day	
2.2.3	Team leader creates development plan	1.0 day	
2.2.4	Development plan circulated to team for agreement	2.0 day	See Appendix V for a template of an example standard test method.
2.2.5	Development plan issued and executed	20.0 day	
2.2.6	Team leader compiles results in a review report/summary	2.0 day	
2.3	Team leader creates/completes draft method	1.0 day	
2.4	Team leader circulates draft method to appropriate team members for review and input	2.0 day	

(Continued)

TABLE 4.1   *(Continued)*

Step	Description	Estimated Duration	Explanation
2.5	Team leader creates draft trial methods validation protocol	2.0 days	The trial methods validation protocol should:
			• Match the appropriate USP method category for use of method • Mirror format and general content of final methods validation protocol • Have acceptance ranges versus acceptance criteria • ID method performance characteristics which are appropriate to the validation
			The trial or practice validation is designed to mimic most of the components of the final validation. Its purpose is to ensure the final validation is executed without deviation, if at all possible. The trial validation protocol and acceptance criteria may be tweaked or modified to create the formal methods validation protocol.
			Note: See Appendix VI for a template of an example methods validation protocol.
2.6	Team members concur on draft method content	1.0 day	
2.7	Team members concur on draft trial methods validation protocol	1.0 day	
2.8	Team leader creates work breakdown structure	1.0 day	
2.9	Team leader creates project plan	1.0 day	Project plan should be a simple document designed to give the trial methods validation process some structure. It may include:
			• Steps to be executed • Personnel assignments • Start and finish dates • Required output for each step • Current status and notes

TABLE 4.1   *(Continued)*

Step	Description	Estimated Duration	Explanation
			Note: It should match/align with the validation SOP and/or protocol.
2.10	Team leader promulgates project plan to team members	1.0 day	
2.11	Changes to project plan?	0.0 day	If yes, team leader makes changes, and promulgates to team members for review.
2.12	Team members approve project plan	0.5 day	
2.13	Team leader solicits feedback from QA on draft trial methods validation protocol	1.0 day	
2.14	Draft method finalized as working method	1.0 day	
2.15	Draft method included in trial validation protocol	0.25 day	
2.16	Trial methods validation protocol is finalized	1.0 day	If practical, the trial validation protocol should give as much detail as possible. This will ensure proper execution of the experiments. Some examples include: • Detailed standard preparation instructions • Detailed sample preparation instructions • Detailed placebo preparation instructions • Detailed spiked placebo preparation instructions • Other clarifying instructions as appropriate
2.17	Team leader performs review of trial methods validation protocol and validation SOP to ensure alignment	0.25 day	
2.18	Trial methods validation protocol is issued and executed	0.25 day	

(Continued)

TABLE 4.1 *(Continued)*

Step	Description	Estimated Duration	Explanation
2.19	Trial methods validation project plan is executed	0.25 day	
2.20	Trial methods validation data collected and reviewed as appropriate	15.0 day	
2.21	Team leader compiles trial methods validation data in draft report format	2.0 day	The purpose of this report is to organize the trial validation data into a final document whose format mimics the desired final formal methods validation report. This is done so that during step 3: formal methods validation and report generation, the final report will require minimum effort to complete. The sections of the report should mirror the trial validation protocol and/or validation SOP. *Note*: Remember, trial methods validation ensures that no surprises occur during formal methods validation. Formal methods validation should be the least complex and time-consuming portion of the validation.
2.22	Compare results obtained to trial validation protocol acceptance criteria ranges	0.5 day	In addition to being part of the body of the report. A table should be generated which shows the desired acceptance criteria, the desired ranges, and the experimentally derived results.
2.23	Team leader promulgates draft trial methods validation report to team members for review and comment	2.0 day	
2.24	Team leader solicits feedback from QA on trial methods validation report format and content	2.0 day	
2.25	Team leader schedules report review meeting with team members and QA	0.5 day	

TABLE 4.1   (*Continued*)

Step	Description	Estimated Duration	Explanation
2.26	Trial validation report review meeting held with validation team members and QA	0.5 day	In the meeting the following items should be addressed: • Were acceptance criteria within the desired ranges? • Did the draft method perform as expected? • Was the protocol lucid and did it capture all instructions needed to execute the trial validation?
2.27	Is the method validatable via existing method and protocol?	0.25 day	If yes, proceed to step 3: Formal methods validation; otherwise continue to step 2.28
2.28	Will modifications require further experiments?	0.25 day	If no, team leader makes appropriate written modifications in method and protocol and proceeds to step 3: formal methods validation; otherwise proceed to step 2.29
2.29	Team identifies modifications needed to protocol and/or draft method which require additional experimentation	0.5 day	An example of this includes forced degradation studies. These are often difficult to control and frequently require additional experimentation to achieve the ~10% to ~30% degradation necessary.
2.30	Abbreviated protocol/work plan generated by team leader	0.5 day	
2.31	Team leader assigns additional laboratory work to team members	0.25 day	
2.32	Team members conduct additional necessary experiments	5.0 day	
2.33	Team leader creates summary report of experiments	2.0 day	
2.34	Team meets and reviews additional experimental results	0.5 day	

(*Continued*)

TABLE 4.1   *(Continued)*

Step	Description	Estimated Duration	Explanation
2.35	Is the method now validatable via method and protocol modification?	0.25 day	If no, return to step 2.29; otherwise continue to step 2.36
2.36	Proceed to step 3: formal methods validation and report generation	0.0 day	
			Total estimated time for step 2 = 75.0 days or 3.8 working months

CHAPTER 5

STEP 3: FORMAL METHODS VALIDATION AND REPORT GENERATION

5.1 BACKGROUND

At the conclusion of the tasks delineated in step 2: final method development and trial methods validation, any ambiguities associated with the method, the methods validation protocol, and the resources and approach taken to perform the validation should no longer exist. The method is now ready to be validated via execution of the methods validation protocol and associated project plan.

It is now possible to meet the FDA's expectation that methods validation is the process of demonstrating that analytical procedure is suitable for its intended use. The methods validation process will now truly be the planned and systematic collection by the applicant of the validation data to support analytical procedures. Validation is also now truly the end game where few surprises and deviations will be encountered. The validation is executed with a formal, approved, and signed methods validation protocol in place which has been reviewed by the quality assurance unit.

At the conclusion of the experimental portion of formal methods validation, a methods validation report, which is supported by the experimental data, is

Validating Chromatographic Methods. By David M. Bliesner
Copyright © 2006 John Wiley & Sons, Inc.

created, circulated, reviewed, and approved by all appropriate personnel. This package is then forwarded to quality assurance for review, approval, archiving, and distribution. Any deviations from the protocol, including failures to meeting acceptance criteria, must be fully investigated and appropriately documented as would any out-of-specification (OOS) result.

Any decision to disregard data, initiate a retest or resample, or modify acceptance criteria must be based on sound scientific data and scientific reasoning. Remember, deviations from the protocol no longer can be arm-waved away. The FDA expects you to scientifically address your failures as you would any other laboratory investigation.

5.2 APPROACH

The basic elements of step 3: formal methods validation and report generation are:

- Draft method is modified as necessary
- Final methods validation protocol is created, reviewed, and approved by all appropriate personnel including quality assurance
- Final project plan is created, reviewed, and approved by all appropriate personnel
- Formal methods validation is executed via implementation of methods validation protocol and project plan
- Data are collected, recorded, and reviewed as appropriate
- Draft methods validation report is created
- Draft methods validation report is reviewed by appropriate personnel
- Draft report and supporting data package are forwarded to quality assurance

Steps in this process are shown in Figure 5.1 and described in Table 5.1.

Components of Step 3: Final Methods Validation and Report Generation
~1 to 2 Months

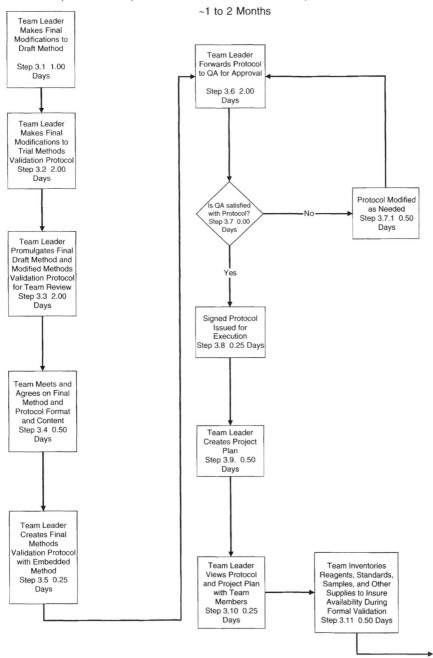

FIGURE 5.1   Workflow diagram for step 3: final methods validation and report generation.

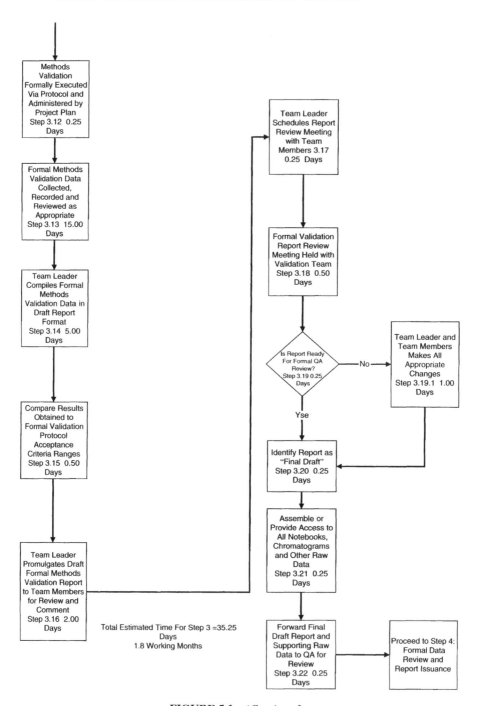

Methods Validation Formally Executed Via Protocol and Administered by Project Plan
Step 3.12 0.25 Days

Formal Methods Validation Data Collected, Recorded and Reviewed as Appropriate
Step 3.13 15.00 Days

Team Leader Compiles Formal Methods Validation Data in Draft Report Format
Step 3.14 5.00 Days

Compare Results Obtained to Formal Validation Protocol Acceptance Criteria Ranges
Step 3.15 0.50 Days

Team Leader Promulgates Draft Formal Methods Validation Report to Team Members for Review and Comment
Step 3.16 2.00 Days

Total Estimated Time For Step 3 =35.25 Days
1.8 Working Months

Team Leader Schedules Report Review Meeting with Team Members 3.17 0.25 Days

Formal Validation Report Review Meeting Held with Validation Team
Step 3.18 0.50 Days

Is Report Ready For Formal QA Review?
Step 3.19 0.25 Days

No →

Team Leader and Team Members Makes All Appropriate Changes
Step 3.19.1 1.00 Days

Yse

Identify Report as "Final Draft"
Step 3.20 0.25 Days

Assemble or Provide Access to All Notebooks, Chromatograms and Other Raw Data
Step 3.21 0.25 Days

Forward Final Draft Report and Supporting Raw Data to QA for Review
Step 3.22 0.25 Days

Proceed to Step 4: Formal Data Review and Report Issuance

FIGURE 5.1   (*Continued*)

**TABLE 5.1   Workflow Diagram for Steps to Step 3:
Final Methods Validation and Report Generation**

Step	Description	Estimated Duration	Explanation
3.1	Team leader makes final modifications to draft method	1.0 day	
3.2	Team leader makes final modifications to trial methods validation protocol	2.0 days	
3.3	Team leader promulgates final draft method and modified methods validation protocol for team review	2.0 days	
3.4	Team meets and agrees on final method and protocol format and content	0.5 day	Acceptance criteria must be set at this point. Criteria should be fully attainable based on results and experience gained from trial validation work.
3.5	Team leader creates final methods validation protocol with embedded method	0.25 day	
3.6	Team leader forwards protocol to QA for approval	2.0 days	
3.7	Is QA satisfied with protocol?	0.0 day	Protocol is modified as needed, and go back to step 3.6.
3.8	Signed protocol issued for execution	0.25 day	
3.9	Team leader creates project plan	0.5 day	Project plan should be a simple document designed to give the trial methods validation process some structure. It may include: • Steps to be executed • Personnel assignments • Start and finish dates • Required output for each step • Current status and notes *Note*: It should match/align with the validation SOP and/or protocol.
3.10	Team leader views protocol and project plan with team members	0.25 day	

(Continued)

TABLE 5.1 *(Continued)*

Step	Description	Estimated Duration	Explanation
3.11	Team inventories reagents, standards, samples, and other supplies to ensure availability during formal validation	0.5 day	
3.12	Methods validation formally executed via protocol and administered by project plan	0.25 day	
3.13	Formal methods validation data collected, recorded, and reviewed as appropriate	15.0 days	All deviations from protocol must be approved and documented appropriately. Deviations should be minimal at this stage.
3.14	Team leader compiles formal methods validation data in draft report format	5.0 days	The trial methods validation report should be used as a template. Data are entered into the report as they are collected and reviewed.
3.15	Compare results obtained to formal validation protocol acceptance criteria ranges	0.5 day	In addition to being part of the body of the report, a table should be generated showing the desired acceptance criteria were met. As stated here, all deviations from protocol must be approved and documented appropriately. Deviations should be minimal at this stage.
3.16	Team leader promulgates draft formal methods validation report to team members for review and comment	2.0 days	
3.17	Team leader schedules report review meeting with team	0.25 day	
3.18	Formal validation report review meeting held with validation team	0.5 day	This should be a critical group discussion. End result should be a report that will stand on its own and be understood by a reviewer who is unfamiliar with the development work. All work presented in the report must have notebook or similar references to raw data. All data tables presented in the report should be reproducible by manual calculation check.
3.19	Is report ready for formal QA review?	0.25 day	

TABLE 5.1 *(Continued)*

Step	Description	Estimated Duration	Explanation
3.20	Identify report as final draft	0.25 day	
3.21	Assemble or provide access to all notebooks, chromatograms, and other raw data	0.25 day	
3.22	Forward final draft report and supporting raw data to QA for review	0.25 day	
	Proceed to step 4: formal data review and report issuance		
			Total estimated time for step 3 = 35.25 days or 1.8 working months.

CHAPTER 6

STEP 4: FORMAL DATA REVIEW AND REPORT ISSUANCE

6.1 BACKGROUND

The process of methods validation is complete when you have:

(1) Demonstrated that you have met all the acceptance criteria
(2) Have clearly documented the results in a CGMP-compliant fashion
(3) Have shown how you met the acceptance criteria in a final methods validation report, including references to raw data, all of which has been reviewed and approved by the appropriate personnel including peers, management, and QA.

Step 4: formal data review and report issuance, is the culmination of the methods validation process. It is a critical component to the process in that:

- It ensures that the reported results are supported by valid scientific data
- The supporting data themselves are corrected as verified by peer and quality assurance personnel review
- It ensures the report is organized in a coherent fashion and can be understood by any reviewer who is reasonably trained in the field of methods validation

Validating Chromatographic Methods. By David M. Bliesner
Copyright © 2006 John Wiley & Sons, Inc.

- There are links to the results reported in the methods validation report so that an outside reviewer can trace the reported results back to raw data
- It ensures the report has received the proper level of review and possesses the corresponding management and QA approval signatures
- The report has been officially issued and controlled copies circulated to the appropriate personnel
- Original copies of the report and all written and electronic data have been properly archived for easy retrieval in the future

Despite its criticality, step 4 is often not considered as a formal part of methods validation by many organizations. Because of this, sufficient resources and emphasis are often not placed on this final component of the methods validation process. This leads to difficulties during review by outside organizations and if difficulties are ever encountered by the end users following technology transfer. Although tedious, proper data review and report issuances procedures are the glue that binds the entire validation process together.

6.2 APPROACH

The basic elements for step 4 are:

- The quality assurance unit (QA) receives all supporting data and the draft methods validation report
- QA performs formal report and data review; corrections made as necessary
- Final methods validation report is generated with appropriate signature cover sheet
- Report is circulated for approval by validation team members, QA, and management
- Controlled copies of final signed methods validation report are made
- QA distributes controlled copies of report as appropriate
- Methods validation report, notebooks, and all raw data are archived as appropriate

Steps in this process are shown in Figure 6.1 and described in Table 6.1. Some details for each step are summarized in Table 6.1.

Formal Data Review and Report Issuance
~1 to 2 Months

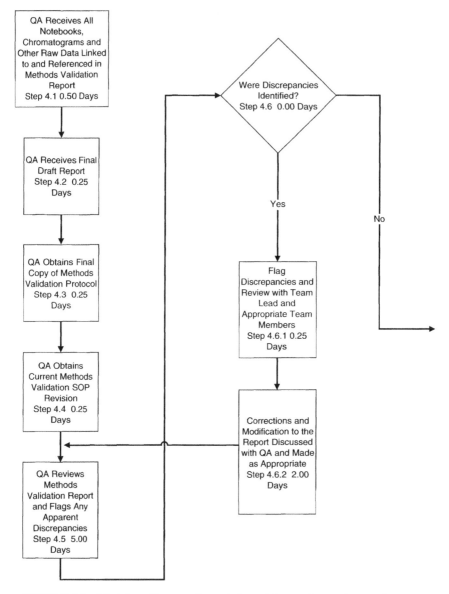

FIGURE 6.1   Workflow diagram of step 4: formal data review and report issuance.

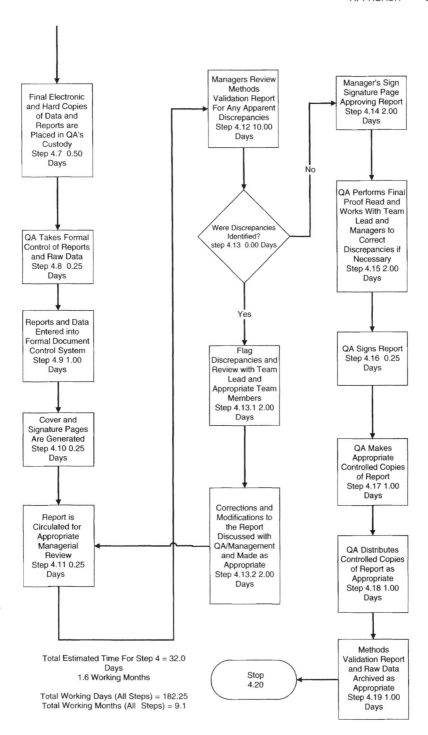

FIGURE 6.1 (*Continued*)

**TABLE 6.1 Workflow Diagram For Steps to Step 4:
Formal Data Review And Reports Issuance**

Step	Description	Estimated Duration	Explanation
4.1	QA receives all notebooks, chromatograms, and other raw data linked to and referenced in methods validation report	0.5 day	
4.2	QA receives final draft report	0.25 day	
4.3	QA obtains final copy of methods validation protocol	0.25 day	
4.4	QA obtains current method validation SOP revision	0.25 day	
4.5	QA reviews method validation report and flags any apparent discrepancies	5.0 days	Report reviewed versus: • Methods validation SOP • Methods validation protocol • Internal checklists • Industry guidance documents Points to consider during report review: • Transcription accuracy verification • Random sampling of notebooks for raw data • Confirmation of table calculations via manual comparison • Training records of all personnel involved in methods validation • Equipment qualification status for instruments used in methods validation • Ability of an external reviewer to read and understand the experiments undertaken (e.g., the report's readability) • Do the data support the claim that the method is validated for its intended use?
4.6	Were discrepancies identified?	0.0 days	If no, then proceed to step 4.7
4.6.1	Flag discrepancies and review with team leader and appropriate team members	0.25 day	

TABLE 6.1 *(Continued)*

Step	Description	Estimated Duration	Explanation
4.6.2	Corrections and modification to the report discussed with QA and made as appropriate	2.0 days	
	Proceed back to step 4.5		
4.7	Final electronic and hard copies of data and reports are placed in QA's custody	0.5 day	
4.8	QA takes formal control of reports and raw data	0.25 day	Notebook entries should have been reviewed by appropriate personnel prior to this point.
4.9	Reports and data entered into formal document control system	1.0 day	
4.10	Cover and signature pages are generated	0.25 day	
4.11	Report is circulated for appropriate managerial review	0.25 day	
4.12	Managers review method validation report for any apparent discrepancies	10.0 days	.
4.13	Were discrepancies identified?	0.0 days	If no, go to step 4.14
4.13.1	Flag discrepancies and review with team leader and appropriate team members	2.0 days	
4.13.2	Corrections and modifications to the report discussed with QA/management and made as appropriate	2.0 days	
	Go back to step 4.11		
4.14	Managers sign signature page approving report	2.0 days	
4.15	QA performs final proofread and works with team leader and managers to correct discrepancies if necessary	2.0 days	
4.16	QA signs report	0.25 day	

(Continued)

TABLE 6.1 *(Continued)*

Step	Description	Estimated Duration	Explanation
4.17	QA makes appropriate controlled copies of report	1.0 day	
4.18	QA distributes controlled copies of report as appropriate	1.0 day	
4.19	Methods validation report and raw data archived as appropriate	1.0 day	
4.20	Stop		Total estimated time For step 4 = ~32.0 days or ~1.6 working months. Total working days (all steps) = ~182.25 Total working months (all steps) = ~ 9.1

CHAPTER 7

SUMMARY

7.1   A BRIEF REVIEW OF THE GUIDE

Over the course of the preceding six chapters, the process of validating a chromatographic method has been discussed in detail. The intent of the guide was to bring order to the potentially chaotic process of methods validation. If you are new to methods validation, we hope the guide provided enough practical information and tools to keep you from having to reinvent the wheel by having to develop your own systems and to attack methods validation from scratch. If you are experienced with methods validation, we hope you found this guide an informative and useful tool that can be used to upgrade and improve existing systems. Although the guide is a practical compilation of documents, it is by no means a technical cookbook on how to validate an analytical method. This would be impossible since every method has its own idiosyncrasies.

In review, the critical features of these discussions within the previous six chapters included:

- An overview of analytical methods validation
- Components of methods validation including method categories and corresponding analytical performance characteristics

Validating Chromatographic Methods. By David M. Bliesner
Copyright © 2006 John Wiley & Sons, Inc.

- The steps required in methods validation which include:
  - Step 1: Method evaluation and further method development
  - Step 2: Final method development and trial methods validation
  - Step 3: Formal methods validation
  - Step 4: Formal data review and report issuance
- Appendices include the following example templates:
  - A glossary of methods validation terms
  - A methods validation SOP
  - An end user requirements questionnaire
  - A method review checklist
  - A standard test method
  - A methods validation protocol
  - A methods validation report

Much of the value of this guide can come from reviewing these documents and modifying them to fit your individual needs. Experienced method validators may take exception to the level of detail, planning, and suggestions in this guide to execute and complete each of the four steps to methods validation. In some cases, their exception may be correct, but overall our experience has shown that the estimates are accurate and reflect the time and effort that should be put forth to properly validate a method.

In closing, it needs to be emphasized that the guide focuses on chromatographic methods validation, specifically high performance liquid chromatographic (HPLC) methods validation. This approach was chosen because HPLC is by far the most common analytical technique used in modern pharmaceutical analytical R&D/QC laboratories. However, the concepts are in many cases directly applicable to validation of other analytical techniques as well.

In addition, it also needs to be emphasized that this guide is specifically written to address the needs and requirements of the pharmaceutical and related industries governed by FDA regulations, specifically the current good manufacturing practice regulations (CGMPs) and international guidance documents such as provided by the International Conference on Harmonization (ICH). However, since the concepts presented in the guide are based on good science, the validation, project management, and documentation practices outlined are in many cases directly applicable to other industries, regulated and nonregulated alike. Also be aware that regulations and documents are always changing and that industry is always improving and upgrading its best practices. Therefore, you are encouraged to keep current by attending conferences, reviewing literature, and networking with your fellow scientists both within and outside your organization.

APPENDIX I

GLOSSARY OF METHODS VALIDATION TERMS

The following terms are commonly encountered while working in a CGMP laboratory during the execution of a methods validation:

483 (Form 483). The designation of the FDA form used to capture and report observations of CGMP deficiencies related to FDA onsite audits. These observations are compiled and become part of the facilities inspection report. Form 483 observations become part of the public record.

Acceptance criteria. Numerical limits, ranges, or other suitable measures used to determine the acceptability of the results of analytical procedures.

Accuracy. Expresses the closeness of agreement between the value found and the value that is accepted as either a conventional true value or an accepted reference value. It may often be expressed as the recovery by the assay of known, added amounts of analyte.

Action level/alert level. Alert level is used to identify the point at which a parameter has drifted toward the extreme of the specified operating range. Action level is when the parameter has drifted outside of the specified operating range. Alert and action levels must be tighter than registration specifications. Alerts are reported to management and evaluated. If an action level is reached, it is reported to management, investigated, and a corrective action initiated.

Validating Chromatographic Methods. By David M. Bliesner
Copyright © 2006 John Wiley & Sons, Inc.

Active pharmaceutical ingredient (API).   Also known as drug substance it is any component that is intended to furnish pharmacological activity or other direct effect in the diagnosis, cure, mitigation, treatment, or prevention of disease, or to affect the structure of any function of the body of man or other animals.

Analytical performance characteristics.   A term used by the USP analytical performance characteristics refer to those characteristics of an analytical method that define its performance as an analytical technique. These performance characteristics include accuracy, precision, specificity, detection limit, quantitation limit, linearity, and range. They need to be considered when validating any one of the USP method categories.

Atypical result.   A result generated on a test article that is within specifications but is inconsistent with previous data, established trends, or other results for the same sample on test.

Audit Summary Report (ASR).   The final output of the laboratory audit. A coherent and organized presentation of findings and suggestions for corrective and preventive actions.

Batch.   A specific quantity of a drug or other material intended to have a uniform character and quality within specified limits and produced according to a single manufacturing order during the same one cycle of manufacture.

Batch record.   A record prepared for each batch of drug product or API produced that includes complete information relating to the production and control of each batch.

Blank.   A sample or standard of a particular matrix or composition without analyte.

Buffer (buffering agent).   A substance or mixture of substances (i.e., phosphate salts) that in solution tend to stabilize the hydrogen ion concentration by neutralizing within limits both acids and bases, so the solution resists changes to pH.

Calibration.   The demonstration that a particular instrument or device produces results within specified limits by comparison with those produced by a reference or traceable standard over an appropriate range of measurements.

Calibration curve.   A plot of standard solution concentration on the x axis versus instrument response on the y axis. In chromatographic analyses, calibration curves are generated by analyzing standard analyte solutions of known concentration and measuring the resulting chromatographic peak area. The resulting plot is then used to determine the concentration of unknown sample solutions containing the same analyte. This is done by measuring the unknown peak area (y) and using the equation for the line to solve for the concentration of the unknown (x). Although referred to as a curve, it is usually a linear plot with a well-defined slope and y intercept.

Capacity factor k'.   A dimensionless quantity used to describe the retention of a compound. It is calculated by the following formula: $(t_r - t_0)/t_0$, where t_r is the measured retention time of the component of interest, and t_0 is the retention time of an unretained component. Retention time t_0 is most measured at the first disturbance of the baseline in HPLC analyses. Retention time t_r is measured at the peak apex. k' is a normalized value for retention. Values range between 2 and 10 for acceptable chromatography.

Change by effect, 30 days (CBE 30).   Supplemental changes to applications that do not require prior approval by FDA. These changes may be implemented within 30 days following submission to the FDA if the agency has no comments.

Change control procedure.   A procedure describing measures to be taken for the purpose of controlling and maintaining an audit trail when changes are made to any part of a system (e.g., standard operating procedure, test method, or specification).

Check standard.   A second preparation of the working standard which is analyzed as part of the system suitability run. The check standard is prepared at the same concentration as the working standard. Prior to continuing the chromatographic run the ratios of the response factors (response factor = area/concentration) for the working standard and the check standard is calculated. $RF_{check\ standard}/RF_{working\ standard}$ should normally be within $\pm$ 2.0%. This provides assurance that the working standard was prepared correctly.

Code of Federal Regulations (CFR).   The codification of the general and permanent rules published in the federal register by the executive departments and agencies of the federal government. It is divided into 50 titles that represent broad areas subject to federal regulation.

Compendial tests methods.   Test methods that appear in official compendia such as the *United States Pharmacopoeia* (USP/NF).

Complaint.   Any verbal, written, or electronic report that alleges deficiencies related to the identity, strength, quality, purity, or effectiveness of a product after it has been released for distribution.

Component.   Any ingredient intended for use in the manufacture of a drug product, including those that may not appear in such drug product and primary packaging components.

Compounding.   Bringing together into homogenous mix of active ingredients, excipients, and, as applicable, solvent components.

Consent decree.   A voluntary legal agreement a drug firm enters into with the FDA for the express purpose of correcting deficiencies related to CGMPs.

Contract research organization (CRO).   A contract manufacturing or analytical testing company.

Corrective action project plan (CAPP).   The project plan generated to ensure successful implementation of corrective and preventive actions.

Corrective and preventive actions (CAPAs).   The steps taken to correct and prevent deficiencies uncovered during a laboratory audit.

Current good manufacturing practices (CGMPs).   21 code of federal regulations, parts 210 and 211. Federal regulations that describe the minimum current good manufacturing practices for preparation of drug products for administration to humans and animals. They include methods to be used in and the facilities or controls to be used for the manufacturing, processing, packing, or holding of a drug to assure that such a drug meets the requirements of the act as to safety and has the identity and strength and meets the quality and purity characteristics that it purports or is represented to possess.

Degradation product.   A molecule resulting from a chemical change in the drug molecule brought about over time and/or the action of light, temperature, pH, water, and so on, or by reaction with an excipient and/or the immediate container/closure system.

Denaturation.   A condition in which a protein unfolds or its polypeptide chains are disordered, rendering the molecule less soluble and usually nonfunctional. The amount of organic modifier in reverse-phase HPLC, if high enough, can denature proteins.

Detection limit.   The detection limit (DL) or limit of detection (LOD) of an individual procedure is the lowest amount of analyte in a sample that can be detected but not necessarily quantitated as an exact value. The LOD is a parameter of limit tests (tests that only determine if the analyte concentration is above or below a specification limit). In analytical procedures that exhibit baseline noise, the LOD can be based on a signal-to-noise ratio (3:1) which is usually expressed as the concentration (e.g., percentage, parts per billion) of analyte in the sample.

Development report.   A report that summarizes the major stages of drug product or API development from early stages through large-scale manufacturing.

Documentation.   Any combination of text, graphics, data, audio, or video information that can be used to clearly and completely recreate an activity, event, or process. Documentation includes life-cycle documents as well as records.

Document control system. A system for managing preparation, review, approval, issuance, distribution, revision, retention, archival, obsolescence, and destruction of life-cycle documents.

Drug product. The combination of API and excipients processed into a dosage form and marketed to the public. Common examples include tablets, capsules, and oral solutions. Also referred to as finished product or dosage form.

Drug substance. See active pharmaceutical ingredient (API).

Effective date. Date by which the approved standard or procedure shall be implemented and in use. All required training must be completed prior to this date.

Excipient(s). A raw material that may perform a variety of roles in a drug product (e.g., tablet press lubricant, filler, diluent, disintegration accelerator, colorant) However, unlike the API, which is pharmacologically active, the excipient has no intrinsic pharmacological activity.

Expiration date. Date placed on the container/labels of a drug product or API designating the time during which a batch of product is expected to remain within the approved shelf life specification if stored under defined conditions, and after which it must not be used.

Extraction efficiency. Measures the effectiveness of extraction of the drug substance from the sample matrix. Studies are conducted during methods validation to determine that the sample preparation scheme is sufficient to ensure complete extraction without being unnecessarily excessive. This is normally investigated by varying the shaking or sonication times (and/or temperature) as appropriate.

Filter study. A comparison of filtered to unfiltered solutions in a methods validation to determine whether the filter being used retains any active compounds or contributes unknown compounds to the analysis.

Food and Drug Administration (FDA). The agency responsible for protecting the public health by assuring the safety, efficacy, and security of human and veterinary drugs, biological products, medical devices, the U.S. food supply, cosmetics, and products that emit radiation. The FDA is also responsible for advancing the public health by helping to speed innovations that make medicines and foods more effective, safer, and more affordable. It helps the U.S. public get accurate, science-based information they need in order to use medicines and foods to improve their health.

Forced degradation studies. Studies undertaken to degrade the sample (e.g., drug product or API) deliberately. These studies, which may be undertaken during method development and/or validation, are used to evaluate an analytical method's ability to measure an active ingredient and its degradation products without interference and are an integral part of the validation of a method as being specific and stability indicating.

Formulation. The recipe describing the quantity and identity of API and excipients making up a drug product. For example, a 100-mg tablet of Advil® may actually weigh 165 mg total. The formulation may include: 100 mg of ibuprofen (the active ingredient), 50 mg of starch (filler), 5 mg of talc (lubricant for the tablet press), and 10 mg of iron oxide (colorant).

Good documentation practices. The handling of written or pictorial information describing, defining, specifying, and/or reporting of certifying activities, requirements, procedures, or results in such a way as to ensure data integrity.

ICH. Tripartite International Conference on Harmonization, an international organization formed to establish uniform guidance within the pharmaceutical industry. For drug development and manufacture, the ICH issues guidance, such as ICH Q2A *Text on Validation of Analytical Procedures* (March 1995), designed to instill uniformity within the industry with respect to various drug development and manufacturing issues. The ICH is composed of industry experts who work jointly and with the FDA to develop these guidance documents.

Identification test. An analytical method capable of determining the presence of the analyte and discriminating between closely related compounds.

In-process control/test. Checks performed during production in order to monitor and, if appropriate, to adjust the process and/or to ensure that the product meets its specifications. The control of the environment or equipment may also be regarded as part of the in-process control.

Installation, Operation, and Performance Qualification (IQ/OQ/PQ). The process by which laboratory equipment is properly installed and determined to be operating within specifications for its intended use. IQ/OQ/PQ is executed via protocol with predetermined acceptance criteria.

Intermediate. Any substance, whether isolated or not, produced by chemical, physical, or biological action at some stage in the production of a drug product or API and subsequently used at another stage of production.

Internal audit. A systematic examination conducted by an internal organizational unit to determine whether quality activities and related results comply with policies, standards, and requirements and whether practices have been implemented effectively and are suitable to achieve objectives and ensure compliance with regulatory requirements.

Label claim. Theoretical strength of the product as given on the marketed product label.

Laboratory audit form (LAF).   The primary data capture instrument used when conducting a laboratory audit.

Laboratory Information Management System (LIMS).   Any computer system used to collect, compile, organize, and report laboratory data. LIMS may also be used to calculate results.

Laboratory investigation report (LIR).   An investigation of any laboratory results or observation which does not meet acceptance criteria or falls outside the expected operational parameters. Similar to an OOS investigation, but does not have to do with meeting a release specification.

Laboratory qualification.   A process where a laboratory that has demonstrated that it has the systems in place necessary to properly perform the tests being conducted.

Linearity.   Evaluates the analytical procedure's ability (within a given range) to obtain a response that is directly proportional to the concentration (amount) of analyte in the sample. If the method is linear, the test results are directly or by well-defined mathematical transformation proportional to the concentration of analyte in samples within a given range. Linearity is usually expressed as the confidence limit around the slope of the regression line. The line is calculated according to an established mathematical relationship from the test response obtained by the analysis of samples with varying concentrations of analyte. Linearity may be established for all active substances, preservatives, and expected impurities. Evaluation is performed on standards. Note that this is different from *range* (sometimes referred to as *linearity of method*), which is evaluated using samples and must encompass the specification range of the component assayed in the drug product.

Lot.   A batch or any portion of a batch having uniform character and quality within specified limits or, in the case of a drug product, is a continuous process; it is a specific identified amount produced in a unit of time or quantity in a manner that assures its having uniform character and quality within specified limits.

Master production record.   A document that specifies complete manufacturing and control instructions, sampling, and testing procedures, specifications, special notations, and precautions to be followed in production of the product and assures uniformity from batch to batch by specifying the batch size, components and quantities, and theoretical yields.

Matrix (sample matrix).   The components and physical form with which the analyte of interest is intimately associated. In the case of drug product, the matrix is the combination of excipients in which the active ingredient is diluted and formed within. For example, the matrix of a transdermal patch is an adhesive in which the drug substance is dissolved, fixed to a plastic backing covered by a release liner. The chemical composition and physical structure of the matrix can have a substantial effect on sample preparation and extraction of the active moiety.

Maximum allowable residue.   Used to calculate the acceptance criteria for cleaning verification methods. Residue limits should be practical, achievable, verifiable, and based on the most deleterious residue. Limits may be established based on the minimum known pharmacological, toxicological, or physiological activity of the API or its most deleterious component.

Method development.   The process by which methods are developed and evaluated for suitability of use as test methods and as precursors to validation.

Method qualification.   Preliminary methods validation conducted during phases 1, 2, and 3 to support the drug development process and the associated release of clinical trial material.

Method transfer.   Moving any analytical technique (chemical or microbiological) from one site/area to another.

Methods validation.   A series of systematic laboratory studies where the performance characteristics of analytical procedures are established to meet the requirements for intended analytical applications. The FDA states in their guidance document that "Methods validation is the process of demonstrating that analytical procedures are suitable for their intended use. The methods validation process for analytical procedures begins with the planned and systematic collection by the applicant of the validation data to support analytical procedures."

Method verification.   The process by which compendial methods are determined to be suitable for analysis of a given test article by a given laboratory. As cited in 21 CFR Part 211.194(a) (2), method verification is used for *USP/NF Compendial* methods or those compendia (and other recognized references such as the *AOAC Book of Methods*) that have documented evidence that the methods have been validated. At a minimum this involves sample solution stability, specificity, and intermediate precision.

Out-of-specification (OOS) investigation.   The systematic and planned search for the root cause that generated an out of specification result. OOS investigations should include formal reporting and a description of corrective and preventative actions taken.

Out-of-specification (OOS) result.   An examination, measurement, or test outcome that does not comply with the specification or predetermined acceptance criteria.

Percent relative standard deviation (%RSD).   A common expression and measure of the relative precision of an analytical method for a given set of measurements. % RSD is calculated by dividing the standard deviation for a series of measurements by the mean of the same sets of measurements and multiplying by 100. % RSD $= (\sigma_{n-1} / \text{mean}) * 100$. Large % RSDs for a series of measurements indicate significant scatter and lack of precision in the technique.

Personnel qualification. The combination of education, training, and/or experience that enables an individual to perform assigned tasks.

Placebo. A formulation containing all ingredients of a drug product except the active ingredient for which the method is being developed.

Preapproval Inspection (PAI). An inspection by the FDA to confirm the CGMP compliance of a drug manufacturing facility. This takes place prior to the FDA's market approval of the drug for sale.

Precision. Expresses the closeness of agreement (degree of scatter) between a series of measurements obtained from multiple aliquots of a homogenous sample under prescribed conditions. The precision of an analytical procedure is usually expressed as the % RSD. According to the FDA and ICH, precision may be considered at three levels, namely:

Repeatability. Refers to the use of the analytical procedure within a laboratory by a single analyst, on a single instrument, under the same operation conditions, over a short interval of time. This is sometimes referred to as *method precision.*

Intermediate precision. Refers to variations within a laboratory as with different days, with different instruments, by different analysts, and so forth. Formally known as *ruggedness.*

Reproducibility. The measure of the capacity of the method to remain unaffected by a variety of conditions such as different laboratories, analysts, instruments, reagent lots, elapsed assay times, and days; more succinctly, the use of the method in different laboratories. Reproducibility is not part of the expected methods validation process. It is addressed during technical transfer of the method to different sites.

Process impurity. Any component of the drug product resulting from the manufacturing process that is not the chemical entity defined as the drug substance or an excipient in the drug product.

Prospective validation. Validation conducted prior to the distribution of either a new product or a product made under a revised manufacturing process where revisions may affect product characteristics.

Protocol. An approved documented experimental design that, when executed, will demonstrate the ability of the subject method to perform as intended. Formal methods validation studies require a protocol. The protocol must have preassigned and approved acceptance criteria for each stage of the validation. The protocol may be a standalone document or may reference the methods validation SOP for specific details.

Q. The amount of dissolved active ingredient specified in the monograph, expressed as a percentage of the labeled content of the dosage form, obtained during dissolution testing. For example, a tablet may have a specification stating that 85% of the active ingredient must be dissolved in 30 minutes of dissolution testing using USP apparatus II (paddles). In this case, $Q = 85\%$. Also, a nefarious character in the *Star Trek: The Next Generation* TV series.

Qualification. Action of proving and documenting that equipment or ancillary systems are properly installed, work correctly, and actually lead to the expected results. Qualification is part of validation, but the individual steps alone do not constitute process validation.

Quality assurance unit (QA). The quality assurance unit serves the role of the quality control unit as defined in 21 CFR 211.11. In this document, only the compliance function of the quality unit is addressed, not the testing functions. In the past there has been some confusion with respect to quality assurance and quality control. CGMPs now generally recognize QA = compliance, QC = testing.

Quality control (QC). The unit responsible for performing testing API and drug product, often referred to as the QC laboratory.

Quality system. A group of interrelated activities representing an integrated approach to philosophy and practices of manufacturing APIs and drug products to assure safety, identity, strength, purity, and quality.

Quality unit. An organizational unit independent of production which ensures that the manufacture, testing, storage, and distribution of drug products, active pharmaceutical ingredients, and components are performed in compliance with regulatory requirements and conformance to company policies and industry practices. Also referred to as quality assurance or QA.

Quantitation limit (QL). Also known as limit of quantitation (LOQ) of an individual analytical procedure, the lowest amount of analyte in a sample that can be quantitatively determined with suitable precision and accuracy. The quantitation limit is a parameter of quantitative assays for low concentrations of compounds in sample matrices and is used particularly for the determination of impurities and/or degradation products. It is usually expressed as the concentration (e.g., percentage, parts per million) of analyte in the sample. For analytical procedures that exhibit baseline noise, the LOQ is generally estimated from a determination of signal-to-noise ratio (10:1) and may be confirmed by experiments.

Quarantine. The status of materials isolated physically or by other effective means to preclude their use pending a decision on their subsequent disposition.

Range. The interval between the upper and lower concentrations (amounts) of analyte in the sample (including these concentrations) for which it has been demonstrated that the analytical procedure has a suitable level of precision, accuracy, and linearity. Range is normally expressed in the same units as test results (e.g., percent, parts per million) obtained by the analytical method. Range (sometimes referred to as *linearity of method*) is evaluated using samples and must encompass the specification range of the component assayed in the drug product.

Raw data. Raw data are defined as the original records of measurement or observation. Raw data may include, but are not limited to, printed instrument output, electronic signal output, computer output, hand-recorded numbers, digital images, hand-drawn diagrams, and so on. Raw data are proof of the original measurement or observation and by definition cannot be regenerated once collected.

Raw material. Any ingredient intended for use in the production of intermediates, APIs, or drug products.

Reference standard. A highly purified compound that is well characterized. It is used as a reference material to confirm the presence and/or amount of the analyte in samples.

Related compounds. Categorized as process impurities, degradants, or contaminants found in finished drug products.

Relative response factor (RRF). The ratio of the response factor of a major component to the response factor of a minor peak. This allows the accurate determination of a minor component without the need for actual standards.

Relative retention time (RRT). The normalization of minor peaks to the parent peak in a chromatogram. The RRT of the parent will be 1.0. Peaks eluting before the parent will have RRTs <1.0. Peaks eluting after the parent will have RRTs >1.0.

Repeatability. The variation experienced by a single analyst on a single instrument. Repeatability does not distinguish between variation from the instrument or system alone and from the sample preparation process.

Reporting limit. The level, at or above the LOQ, below which values are not reported (e.g., reported as < 0.05% for a reporting limit = 0.05%). The reporting limit may be defined by ICH thresholds.

Reprocessing. Introducing a previously processed material, which does not conform to standards or specifications, back into the process and repeating a step or steps that are part of the established manufacturing process.

Resolution. A measure of the efficiency of the separation of two component mixtures. In chromatographic analyses a resolution of >1.5, which means two peaks are separated from each other all the way to the baseline, is desirable.

Response factor. A measuring of the signal generated by a detector normalized to the amount of analyte present. In HPLC, it is usually the peak area for a given component (the response) divided by the concentration or mass that generated that response.

Retain sample. Reserve samples. An appropriately identified sample that is representative of each lot or batch of drug product or API stored under conditions consistent with product labeling, in the same container–closure system in which the product is marketed or in one that has essentially the same characteristics. Used for testing that may be involved in analyzing complaint samples.

Retrospective validation. Validation of a process for a product already in distribution based upon accumulated production, testing, and control data.

Revalidation. The process of partially or completely validating a method or process after changes or modifications have been made to the manufacturing process, analytical methodology, equipment, instrumentation, or other parameter(s) that may affect the quality and composition of the finished product. The USP specifically sites changes in synthesis of drug substance, changes in the composition of the drug product, and changes in analytical procedure.

Robustness. The measure of the ability of an analytical method to remain unaffected by small but deliberate variations in method parameters (e.g., pH, mobile phase composition, temperature, instrument settings); provides an indication of its reliability during normal usage. Robustness testing is a systematic process of varying a parameter and measuring the effect on the method by monitoring system suitability and/or the analysis of samples. It is part of the formal methods validation process.

Ruggedness. A dated term now commonly accepted as intermediate precision. See definition of **intermediate precision.**

Selectivity. The ability of the method to separate the analyte from other components that may be present in the sample, including impurities. Determination of selectivity normally includes analyzing placebo, blank, media for dissolution, dilution solvent, and mobile phase injections. Also, no chromatographic peaks, such as related compounds, should interfere with the analyte peak or internal standard peak, if applicable. Selectivity=separate and show every component in the sample.

Signal to Noise (S/N). In chromatography the measure of average baseline noise (e.g., peak-to-peak) to the signal given by an analyte peak. S/N calculations are performed when determining the LOD and LOQ.

Specification. The quality standards (e.g., tests, analytical procedures, and acceptance criteria) provided in an approved application to confirm the quality of drug substances, drug products, intermediates, raw materials, reagents, and other components including in-process materials.

Specificity. The ability to assess unequivocally the analyte in the presence of components that may be expected to be present such as impurities, degradation products, and excipients. There must be inarguable data for a method to be specific. Specificity=measures only the desired component without interference from other species that might be present; separation is not necessarily required.

Spiked placebo. Preparation of a sample to which known quantities of analyte are added to placebo material. Performed during validation to generate accurate and reproducible samples used to demonstrate recovery from the sample matrix.

Spiking. The addition of known amounts of a known compound to a standard, sample, or placebo typically for the purpose of confirming the performance of an analytical procedure or the calibration of an instrument.

Stability indicating methodology. A validated quantitative analytical procedure or set of procedures that can detect the changes with time in the pertinent properties (e.g., active ingredient, preservative level, or appearance of degradation products) of the drug substance and drug product

Stability indicating assay. An assay that accurately measures the component of interest [the active ingredient(s) or degradation products] without interference from other degradation products, process impurities, excipients, or other potential interfering substances.

Stability indication profile. A set of procedures or assays that collectively detect changes with time, although may not do so individually.

Standard and sample solution stability. Established under normal benchtop conditions, normal storage conditions, and sometimes in the instrument (e.g., an HPLC autosampler) to determine if special storage conditions are necessary, for instance, refrigeration or protection from light. Stability is determined by comparing the response and impurity profile from aged standards or samples to that of a freshly prepared standard and to its own response from earlier time points. Note that these are short-term studies and are not intended to be part of the stability indication assessment or product stability program.

Stressed studies. *See* **Forced degradation studies.**

Subject matter expert (SME). An individual who is consider to be an expert on a particular subject due to a combination of education, training, and experience.

System suitability. Evaluation of the components of an analytical system to show that the performance of a system meets the standards required by a method. A system suitability evaluation usually contains its own set of parameters. For chromatographic assays, these may include tailing factors, resolution, and precision of standard peak areas, and comparison to a confirmation standard, capacity factors, retention times, theoretical plates, and calibration curve linearity.

Tailing factor. A measure of peak asymmetry. Peaks with a tailing factor of >2 are usually considered to be unacceptable due to difficulties in determine peak start and stop points which complicates integration. Tailing peaks are an indication that the chromatographic conditions for a separation have not been properly optimized.

Technology transfer. The transfer of a process or method, including demonstration of equivalence according to predetermined criteria, between the receiving and transferring sites.

Test method. An approved, detailed procedure describing how to test a sample for a specified attribute (e.g., assay), including the amount required, instrumentation, reagents, sample preparation steps, data generation steps, and calculations use for evaluation.

Theoretical plates. A dimensionless quantity used to express the efficiency or performance of a column under specific conditions. A decrease in theoretical plates can be an indication of HPLC column deterioration.

Transcription accuracy verification (TAV). The process where the transcription of data from one location to another is confirmed by a second party. Important with methods validation reports with summary results, which are often transcribed and not linked to raw data.

USP <1225> Category I. One of four method categories for which validation data should be required. Category I methods include analytical methods for quantitation of major components of bulk drug substances or active ingredients (including preservatives) in finished pharmaceutical products. Category I methods are typically referred to as assays.

USP <1225> Category II. One of four method categories for which validation data should be required. Category II methods include analytical methods for determination of **impurities** in bulk drug substances or **degradation compounds** in finished pharmaceutical products. These methods include quantitative assays and limits tests.

USP <1225> Category III. One of four method categories for which validation data should be required. Category III methods include analytical methods for determination of performance characteristics (e.g., **dissolution**, drug release).

USP <1225> Category IV. One of four method categories for which validation data should be required. Category IV methods include analytical methods used as **identification tests**.

Validation.   A documented program that provides a high degree of assurance that a specific process, method, or system will consistently produce a result meeting predetermined acceptance criteria.

Validation characteristics.   *See* Analytical performance characteristics.

Validation parameters.   *See* Analytical performance characteristics.

Validation protocol.   A prospective plan that, when executed as intended, produces documented evidence that a process, method, or system has been properly validated.

Validation report.   A summary of experiments and results that demonstrate the method is suitable for its intended use, approved by responsible parties.

Verification.   The act of reviewing, inspecting, testing, checking, or otherwise establishing and documenting whether items, processes, services, or documents conform to specified requirements.

APPENDIX II

TEMPLATE FOR AN EXAMPLE METHODS VALIDATION STANDARD OPERATING PROCEDURE (SOP)

SOP EXAMPLE TEMPLATE

1. PURPOSE

1.1 This procedure is intended to provide general guidelines for the validation of chromatographic methods for the analysis of drug substance and drug product.

1.2 This procedure is also applicable to fields outside the pharmaceutical industry in that it is based on sound scientific principles and broadly recognized practices.

1.3 This procedure is primarily intended for validation of high-performance liquid chromatographic (HPLC) methods but is applicable, with modification, to other chromatographic techniques.

2. SCOPE

2.1. This procedure is applicable to any analytical laboratory involved in the research, development, and validation of chromatographic methods used to ensure the identity and measure the strength, quality, purity, and potency of active pharmaceutical ingredients and drug products.

Validating Chromatographic Methods. By David M. Bliesner
Copyright © 2006 John Wiley & Sons, Inc.

2.2. This procedure is in alignment with current industry practice and current ICH and FDA guidelines.

2.3. If the methods validation protocol differs in its requirements compared to this SOP then the validation protocol takes precedence over this SOP.

3. RESPONSIBILITIES

3.1 Responsibilities for validating analytical methods within an analytical laboratory vary from organization to organization.

3.2. Regardless of the organization, certain roles and responsibilities are generally applicable.

3.3. In particular, the **front-line supervisor** is responsible for ensuring the following:

3.3.1. The need for validation, the validation parameters and requirements to be included, and the acceptance criteria are properly determined.

3.3.2. The generation of a draft methods validation protocol which is then circulated to pertinent parties (e.g., subject matter experts) within the organization for review.

3.3.3. The protocol is revised per comments and additional review is conducted if there are major technical changes.

3.3.4. All personnel have the proper combination of education, training, and experience for the tasks assigned, and that these data are properly documented.

3.3.5. The validation tasks are explained to the analyst(s) including: purpose of the method, parameters, equipment, procedures, criteria, timeline, and end users.

3.3.6. Methods validation is performed as per current industry guidelines cited in this SOP.

3.3.7. The validation study documentation has been reviewed and approved.

3.3.8. The validation protocol is signed and dated.

3.3.9. Methods validation is performed as per the validation protocol and that any and all deviations are appropriately documented.

3.3.10. Ensure that the results for each experiment are reviewed in a timely manner, either by reviewing the results personally or by assigning the review to a qualified analyst.

3.3.11. Meet with the analyst(s) periodically to discuss validation progress and deviations or nonconforming results.

3.3.12. Evaluate and respond to reports of validation study deviations and nonconforming results.

3.3.13. Review the validation report for consistency with the experimental design and protocol, completeness of the descriptions, and correctness of the interpretation of the results.

3.4. The **analyst(s)** will:

3.4.1. Execute the experiments in accordance with the protocol and following CGMPs.

3.4.2. Use equipment that is properly qualified and calibrated.

3.4.3. Use equipment within its designed operating parameters.

3.4.4. Use standards that are within expiration and are properly qualified according to current industry practice. This does not apply to samples since expired samples are frequently used during methods validation.

3.4.5. Promptly record and properly document all data.

3.4.6. Report all deviations from the study or protocol including nonconforming results to the supervisor.

3.4.7. Provide the data reviewers (e.g., second analyst or supervisor) with all notebooks, records, and data needed to evaluate the results.

3.4.8. Retain all standards, samples, preparations, reagents, and glassware used for a given validation study until data have been reviewed by the second analyst or supervisor and the validation report has been signed, if possible.

3.4.9. Maintain all equipment in "as-used" condition until data have been reviewed, if possible.

3.4.10. Assist in preparation of the draft validation report.

3.4.11. At a minimum, review the sections of the validation report containing their work, confirming the accuracy of the data and proper description of the experiments and deviations.

3.5. **Second-level supervisor** is responsible for:

3.5.1 Ensuring the type of validation requirements and experiments conducted are appropriate for final application of the method.

3.5.2 Ensuring the validation report is completed and issued in a timely fashion.

3.5.3. Reviewing and approving the final validation report to confirm that the analytical method is suitable for its intended use.

3.5.4. Ensuring that the quality assurance unit was involved in the review and approval of the final report.

3.6. The **quality assurance unit** is responsible for:

3.6.1. Reviewing and approving the validation protocol to ensure it includes appropriate experimental design and acceptance criteria.

3.6.2. Ensuring that all personnel involved in the validation study were properly trained on all applicable procedures and techniques and possess and have the proper combination of education, training, and experience for the tasks assigned, and that these data are properly documented.

3.6.3. At a minimum randomly sampling and reviewing raw data (e.g., notebooks, chromatograms) associated and referenced in the methods validation report to confirm the integrity of the data reported.

3.6.4. Confirming the integrity of the data presented in the validation report by reproducing calculations and plots presented in the methods validation report without referring to original data.

3.6.5. Verifying transcription of raw data presented in the validation report.

3.6.6. Reviewing and approving the final validation report to confirm that the protocol was properly executed and that any deviations have been addressed and documented.

3.7. Personnel associated with **document control** will:

3.7.1. Ensure that protocols and reports are formally entered into the document control system.

3.7.2. Make appropriate numbers of controlled copies of the methods validation report.

3.7.3. Ensure that the protocol, report, and raw data are properly archived.

3.7.4. Ensure that the protocol, report, and raw data can be accessed and retrieved within a reasonable time frame such as would be required if requested by an FDA auditor during a pre-approval inspection.

4. DEFINITIONS

4.1. Definitions of terms commonly encountered when discussing analytical methods validation are included in Attachment I: Glossary.

5. PROCEDURE

5.1. This section makes some suggestions on what general administrative steps should be taken to validate chromatographic methods.

    5.1.1. The specific steps to be taken are dependent upon the USP method category under which the method to be validated falls. Attachment II: USP Method Categories and Data Elements Required for Validation defines these categories and delineates which analytical performance characteristics and additional validation testing need to be evaluated for each category during methods validation.

    5.1.2. Specific experimental steps and associated recommendations for acceptance criteria for each type of method category are listed in Attachments III–VII.

5.2. The recommended general administrative steps to take when validating method are:

    5.2.1. Working with the second-level supervisor and analyst(s), the front-line supervisor will:

        5.2.1.1. Determine the need for methods validation.

        5.2.1.2. Initiate validation after receiving formal written correspondence from the second-level supervisor.

        5.2.1.3. Evaluate the method with respect to its intended use and determine the method category (see Attachment II: USP Method Categories and Data Elements Required for Validation).

        5.2.1.4. Determine if the method is intended to be stability indicating.

        5.2.1.5. Identify those analytical performance characteristics required for validation.

        5.2.1.6. Review the recommended experimental procedures and acceptance criteria for the appropriate method category as outlined in Attachments III–VI.

        5.2.1.7. Obtain and review all method development reports and documentation.

        5.2.1.8. If method development work was performed outside the validating laboratory, the validating laboratory should perform some prevalidation studies to gain familiarity with method performance.

            5.2.1.8.1. These prevalidation studies should concentrate on assessment of the relevant method performance parameters and the impact on the suitability of the method for its intended use.

5.2.1.8.2. Only after prevalidation studies have been fully completed should acceptance criteria be established to perform full validation.

5.2.1.9. Once the decision has been made to begin the validation experiments, a validation protocol will be generated by the front-line supervisor.

5.2.1.9.1. The validation protocol should follow the recommendations made in the SOP attachments with appropriate modification of experiments and acceptance criteria as necessary.

5.2.1.9.2. The protocol should reference this SOP.

5.2.1.9.3. The protocol should be reviewed by the analysts for content and corrections as needed.

5.2.1.10. Once completed, the protocol is to be forwarded to the second-level supervisor and quality assurance for approval and inclusion into the document control system.

5.2.1.11. After approval the front-line supervisor and analysts execute the protocol and begin collecting data.

5.2.1.12. As the data are being generated, the front-line supervisor along with the analyst review the data, making corrections as necessary.

5.2.1.12.1. Any deviations from the protocol must be formally documented and scientifically justified.

5.2.1.12.2. Quality assurance should be involved with reviewing and capturing these deviations.

5.2.1.13. As the data are being reviewed and approved the front-line supervisor begins drafting the methods validation report.

5.2.1.14. Once the data are collected and reviewed, the front-line supervisor will complete the draft methods validation report and circulate it among the analysts for review and correction as necessary.

5.2.1.15. At the completion of the review, the front-line supervisor will circulate the draft report to the second-level supervisor for review.

5.2.1.16. After the front-line supervisor's comments and corrections are addressed, the draft report is forwarded to quality assurance for review.

5.2.1.17. After quality assurance's comments and corrections are addressed, the final report is issued for approval signatures.

5.2.1.18. After approval the final report and all raw data are forwarded to quality assurance for inclusion into the document's control system.

5.2.1.19. Controlled copies of the validation report are made and circulated to appropriate personnel as needed.

6. METHODS VALIDATION REPORT ELEMENTS

6.1. The methods validation report should contain the following components:

6.1.1. Title

6.1.1.1. Title should give full description of validation experiment undertaken including technique/method validate and product description.

6.1.1.2. Cover page:

6.1.1.2.1. Report number (formal document control number)

6.1.1.2.2. Protocol number

6.1.1.2.3. Month, day, and year issued

6.1.2. Table of contents delineating report contents which should include:

6.1.2.1. Quality assurance review signature page

6.1.2.2. Methods validation report approval signature page

6.1.2.3. Introduction

6.1.2.4. Purpose

6.1.2.5. Experimental

6.1.2.6. Results and discussion

6.1.2.7. Conclusion

6.1.2.8. Figures

6.1.2.9. Analytical method

6.1.2.10. Protocol

6.1.2.11. Protocol deviations

6.1.3. The body of the report, which includes the sections delineated in the table of contents.

6.2. The experimental section should mirror the format of the methods validation protocol. It should include:

6.2.1. A brief description of the experiments and how they were conducted.

6.2.2. A statement of results including reference to tables and figures as appropriate.

6.2.3. A statement comparing the results found and if they match the protocol stated acceptance criteria.

6.2.4. A brief statement if there were deviations from the protocol.

6.2.5. A reference to the location of raw data.

6.3. The results and discussion section should summarize whether the method was successfully validated for its intended use.

6.3.1. This section should address any potential difficulties encountered during validation.

6.3.2. It should discuss any deviations from the protocol.

6.4. The conclusions section should address any limitations of the method and special considerations needed during its implementation.

6.5. The figures section should show linearity and residuals plots as well as example chromatograms supporting the conclusion that the method is validated.

6.6. The remaining sections are presented at face value.

6.7. The methods validation report is not a development report but a summary report.

6.7.1. The purpose of the report is to prove to an outside observer that the method is appropriate for its intended use.

6.7.2. Enough data and description should be present so that the report stands on its own from this perspective; however, it should not be so overly detailed as to be burdensome to the reader.

6.7.3 The reader should not be inclined to want to go to raw data after reviewing the report, but the report and documentation system should make this process effortless if so desired.

7. REFERENCES

7.1. Code of Federal Regulations, Food and Drugs, Title 21 Part 211 "Current Good Manufacturing Practices of Finished Pharmaceuticals"

7.2. Guidance for Industry: *Analytical Procedures and Methods Validation* (Draft, August 2000)

7.3. ICH Q2A: *Text on Validation of Analytical Procedures* (March 1995)

7.4. ICH Q2B: *Validation of Analytical Procedures: Methodology* (May 1997)

7.5. Chapter <621> Chromatography; *US Pharmacopoeia* (current), United States Pharmacopoeia Convention, Inc. Rockville, MD

7.6. Chapter <1225> Validation of Compendial Procedures; *US Pharmacopoeia* (current), United States Pharmacopoeia Convention, Inc. Rockville, MD

7.7. AOAC *Book of Methods* (current)

8. ATTACHMENTS

8.1. Attachment I: Glossary of Terms

8.2. Attachment II: USP Method Categories and Analytical Performance Characteristics Evaluation and Testing Required for Validation

8.3. Attachment III: Recommended Experimental Procedures and Acceptance Criteria: Category I Methods (Assay)

8.4. Attachment IV: Recommended Experimental Procedures and Acceptance Criteria: Category II Methods (Impurity and Degradants)

8.5. Attachment V: Recommended Experimental Procedures and Acceptance Criteria: Category III Methods (Dissolution)

8.6. Attachment VI: Recommended Experimental Procedures and Acceptance Criteria: Category IV Methods (Identification)

ATTACHMENT I: GLOSSARY OF TERMS

The following terms are commonly associated with methods validation documentation:

Acceptance criteria. Numerical limits, ranges, or other suitable measures used to determine the acceptability of the results of analytical procedures.

Accuracy. Expresses the closeness of agreement between the value found and the value that is accepted as either a conventional true value or an accepted reference value. It may often be expressed as the recovery by the assay of known, added amounts of analyte.

Active pharmaceutical ingredient (API). Also known as drug substance, it is component that is intended to furnish pharmacological activity or other direct effect in the diagnosis, cure, mitigation, treatment, or prevention of disease, or to affect the structure of any function of the body of man or other animals.

Analytical performance characteristics. A term used by the USP, analytical performance characteristics refer to those characteristics of an analytical method that define its performance as an analytical technique. These performance characteristics include accuracy, precision, specificity, detection limit, quantitation limit, linearity, and range. They need to be considered when validating any one of the USP method categories.

Blank. A sample or standard of a particular matrix or composition without analyte.

Calibration curve. A plot of standard solution concentration, on the x axis, versus instrument response, on the y axis. In chromatographic analyses, calibration curves are generated by analyzing standard analyte solutions of known concentration and measuring the resulting chromatographic peak area. The resulting plot is then used to determine the concentration of unknown sample solutions containing the same analyte. This is done by measuring the unknown peak area (y) and using the equation for the line to solve for the concentration of the unknown (x). Although referred to as a curve, it is usually a linear plot with a well-defined slope and y intercept.

Capacity factor k'. A dimensionless quantity used to describe the retention of a compound. It is calculated by the following formula: $(t_r - t_0)/t_0$, where t_r is the measured retention time of the component of interest, and t_0 is the retention time of an unretained component. Retention time t_0 is most measured at the first disturbance of the baseline in HPLC analyses. Retention time t_r is measured at the peak apex. k' is a normalized value for retention. Values range between 2 and 10 for acceptable chromatography.

Change control procedure. A procedure describing measures to be taken for the purpose of controlling and maintaining an audit trail when changes are made to any part of a system (e.g., standard operating procedure, test method or specification).

Check standard. A second preparation of the working standard which is analyzed as part of the system suitability run. The check standard is prepared at the same concentration as the working standard. Prior to continuing the chromatographic run the ratios of the response factors (response factor = area/concentration) for the working standard and the check standard is calculated. $RF_{check\ standard}/RF_{working\ standard}$ should normally be within $\pm\ 2.0\%$. This provides assurance that the working standard was prepared correctly.

Compendial tests methods. Test methods that appear in official compendia such as the United States Pharmacopoeia (USP/NF).

Degradation product. A molecule resulting from a chemical change in the drug molecule brought about over time and/or the action of light, temperature, pH, water, and soon, or by reaction with an excipient and/or the immediate container/closure system.

Detection limit. The detection limit (DL) or limit of detection (LOD) of an individual procedure is the lowest amount of analyte in a sample that can be detected but not necessarily quantitated as an exact value. The LOD is a parameter of limit tests (tests that only determine if the analyte concentration is above or below a specification limit). In analytical procedures that exhibit baseline noise, the LOD can be based on asignal-to-noise ratio (3:1),which is usually expressed as the concentration (e.g., percentage, parts per billion) of analyte in the sample.

Drug product. The combination of API and excipients processed into a dosage form and marketed to the public. Common examples include tablets, capsules, and oral solutions. Also referred to as finished product or dosage form.

Drug substance. See active pharmaceutical ingredient (API).

Excipient(s). A raw material that may perform a variety of roles in a drug product (e.g., tablet press lubricant, filler, diluent, disintegration accelerator, colorant) However, unlike the API, which is pharmacologically active, the excipient has no intrinsic pharmacological activity.

Extraction efficiency. Measures the effectiveness of extraction of the drug substance from the sample matrix. Studies are conducted during methods validation to determine that the sample preparation scheme is sufficient to ensure complete extraction without being unnecessarily excessive. This is normally investigated by varying the shaking or sonication times (and/or temperature) as appropriate.

Filter study. A comparison of filtered to unfiltered solutions in a methods validation to determine whether the filter being using retains any active compounds or contributes unknown compounds to the analysis.

Forced degradation studies. Studies undertaken to degrade the sample (e.g., drug product or API) deliberately. These studies, which may be undertaken during method development and/or validation are used to evaluate an analytical method's ability to measure an active ingredient and its degradation products without interference and are an integral part of the validation of a method as being specific and stability indicating.

Formulation. The recipe describing the quantity and identity of API and excipients making up a drug product. For example, a 100-mg tablet of Advil® may actually weigh 165 mg total. The formulation may include: 100 mg of ibuprofen (the active ingredient), 50 mg of starch (filler), 5 mg of talc (lubricant for the tablet press), and 10 mg of iron oxide (colorant).

ICH. The Tripartite International Conference on Harmonization, an international organization formed to establish uniform guidance within the pharmaceutical industry. For drug development and manufacture, the ICH issues guidance, such as ICH Q2A *Text on Validation of Analytical Procedures* (March 1995), designed to instill uniformity within the industry with respect to various drug development and manufacturing issues. The ICH is composed of industry experts who work jointly and with the FDA to develop these guidance documents.

Identification test. An analytical method capable of determining the presence of the analyte and discriminating between closely related compounds.

Label claim. Theoretical strength of the product as given on the marketed product label.

Laboratory qualification. A process where a laboratory that has demonstrated that it has the systems in place necessary to properly perform the tests being conducted.

Linearity. Evaluates the analytical procedure's ability (within a give range) to obtain a response that is directly proportional to the concentration (amount) of analyte in the sample. If the method is linear, the test results are directly or by well-defined mathematical transformation proportional to the concentration of analyte in samples within a given range. Linearity is usually expressed as the confidence limit around the slope of the regression line. The line is calculated according to an established mathematical relationship from the test response obtained by the analysis of samples with varying concentrations of analyte. Linearity may be established for all active substances, preservatives, and expected impurities. Evaluation is performed on standards. Note that this is different from *range* (sometimes referred to as *linearity of method*), which is evaluated using samples and must encompass the specification range of the component assayed in the drug product.

Matrix (sample matrix). The components and physical form with which the analyte of interest is intimately associated. In the case of drug product, the matrix is the combination of excipients in which the active ingredient is diluted and formed within. For example, the matrix of a transdermal patch is an adhesive in which the drug substance is dissolved, fixed to a plastic backing covered by a release liner. The chemical composition and physical structure of the matrix can have a substantial effect on sample preparation and extraction of the active moiety.

Maximum allowable residue. Used to calculate the acceptance criteria for cleaning verification methods. Residue limits should be practical, achievable verifiable and based on the most deleterious residue. Limits may be established based on the minimum known pharmacological, toxicological, or physiological activity of the API or its most deleterious component.

Method development. The process by which methods are developed and evaluated for suitability of use as test methods, and as precursors to validation.

Method qualification. Preliminary methods validation conducted during phases 1, 2, and 3 to support the drug development process and the associated release of clinical trial material.

Method transfer. Moving any analytical technique (chemical or microbiological) from one site/area to another.

Methods validation. A series of systematic laboratory studies where the performance characteristics of the analytical method meet the requirements for intended analytical applications. The FDA states in their guidance document that "Methods validation is the process of demonstrating that analytical procedures are suitable for their intended use. The methods validation process for analytical procedures begins with the planned and systematic collection by the applicant of the validation data to support analytical procedures."

Method verification. The process by which compendial methods are determined to be suitable for analysis of a given test article by a given laboratory. As cited in 21 CFR Part 211.194(a) (2), method verification is used for *USP/NF compendial* methods or those compendia (and other recognized references such as the *AOAC Book of Methods*) that have documented evidence that the methods have been validated. At a minimum this involves sample solution stability, specificity, and intermediate precision.

Percent relative standard deviation (% RSD). A common expression and measure of the relative precision of an analytical method for a given set of measurements. % RSD is calculated by dividing the standard deviation for a series of measurements by the mean of the same sets of measurements and multiplying by 100. % RSD = $(\sigma^{n-1} / \text{mean}) * 100$. Large % RSDs for a series of measurements indicate significant scatter and lack of precision in the technique.

Placebo. A formulation containing all ingredients of a drug product except the active ingredient for which the method is being developed.

Precision. Expresses the closeness of agreement (degree of scatter) between a series of measurements obtained from multiple aliquots of a homogenous sample under prescribed conditions. The precision of an analytical procedure is usually expressed as the % RSD. According to the FDA and ICH, precision may be considered at three levels, namely:

Repeatability. Refers to the use of the analytical procedure within a laboratory by a single analyst, on a single instrument, under the same operation conditions, over a short interval of time. This is sometimes referred to as *method precision.*

Intermediate precision. Refers to variations within a laboratory as with different days, with different instruments, by different analysts, and so forth. Formally known as ruggedness.

Reproducibility. Is the measure of the capacity of the method to remain unaffected by a variety of conditions such as different laboratories, analysts, instruments, reagent lots, elapsed assay times, and days; more succinctly: the use of the method in different laboratories. Reproducibility is not part of the expected methods validation process. It is addressed during technical transfer of the method to different sites.

Process impurity. Any component of the drug product resulting from the manufacturing process that is not the chemical entity defined as the drug substance or an excipient in the drug product.

Protocol. An approved documented experimental design that, when executed, will demonstrate the ability of the subject method to perform as intended. Formal methods validation studies require a protocol. The protocol must have preassigned and approved acceptance criteria for each stage of the validation. The protocol may be a standalone document or may reference the methods validation SOP for specific details.

Q. The amount of dissolved active ingredient specified in the monograph, expressed as a percentage of the labeled content of the dosage form, obtained during dissolution testing. For example, a tablet may have a specification stating that 85% of the active ingredient must be dissolved in 30 minutes of dissolution testing using USP apparatus II (paddles). In this case, $Q = 85\%$. Also, a nefarious character in the *Star Trek: The Next Generation* TV series.

Quality assurance unit (QA). The quality assurance unit serves the role of the quality control unit as defined in 21 CFR 211.11. In this document, only the compliance function of the quality unit is addressed, not the testing functions. In the past there has been some confusion with respect to quality assurance and quality control. CGMPs now generally recognize QA = compliance, QC = testing.

Quantitation limit. Also known as limit of quantitation (LOQ) of an individual analytical procedure, the lowest amount of analyte in a sample that can be quantitatively determined with suitable precision and accuracy. The quantitation limit is a parameter of quantitative assays for low concentrations of compounds in sample matrices and is used particularly for the determination of impurities and/or degradation products. It is usually expressed as the concentration (e.g., percentage, parts per million) of analyte in the sample. For analytical procedures that exhibit baseline noise the LOQ is generally estimated from a determination of signal-to-noise ratio (10:1) and may be confirmed by experiments.

Range. The interval between the upper and lower concentrations (amounts) of analyte in the sample (including these concentrations) for which it has been demonstrated that the analytical procedure has a suitable level of precision, accuracy and linearity. Range is normally expressed in the same units as test results (e.g., percent, parts per million) obtained by the analytical method. Range (sometimes referred to as *linearity of method*) is evaluated using samples and must encompass the specification range of the component assayed in the drug product.

Raw data. Raw data are defined as the original records of measurement or observation. Raw data may include, but are not limited to, printed instrument output, electronic signal output, computer output, hand-recorded numbers, digital images, hand-drawn diagrams, and so on. Raw data are proof of the original measurement or observation and by definition cannot be regenerated once collected.

Reference standard. A highly purified compound that is well characterized. It is used as a reference material to confirm the presence and/or amount of the analyte in samples.

Related compounds. Categorized as process impurities, degradants, or contaminants found in finished drug products.

Relative response factor (RRF). The ratio of the response factor of a major component to the response factor of a minor peak. This allows the accurate determination of a minor component without the need for actual standards.

Relative retention time (RRT). The normalization of minor peaks to the parent peak in a chromatogram. The RRT of the parent will be 1.0. Peaks eluting before the parent will have RRTs < 1.0. Peaks eluting after the parent will have RRTs > 1.0.

Repeatability. The variation experienced by a single analyst on a single instrument. Repeatability does not distinguish between variation from the instrument or system alone and from the sample preparation process.

Reporting Limit. The level, at or above the LOQ, below which values are not reported (e.g., reported as < 0.05% for a reporting limit = 0.05%). The reporting limit may be defined by ICH thresholds.

Resolution. A measure of the efficiency of the separation of two component mixtures. In chromatographic analyses a resolution of > 1.5, which means two peaks are separated from each other all the way to the baseline, is desirable.

Response factor. A measuring of the signal generated by a detector normalized to the amount of analyte present. In HPLC, it is usually the peak area for a given component (the response) divided by the concentration or mass that generated that response.

Revalidation. The process of partially or completely validating a method or process after changes or modifications have been made to the manufacturing process, analytical methodology, equipment, instrumentation, or other parameter(s) that may affect the quality and composition of the finished product. The USP specifically sites changes in synthesis of drug substance; changes in the composition of the drug product; and changes in analytical procedure.

Robustness. The measure of the ability of an analytical method to remain unaffected by small but deliberate variations in method parameters (e.g., pH, mobile phase composition, temperature, instrument settings); provides and indication of its reliability during normal usage. Robustness testing is a systematic process of varying a parameter and measuring the effect on the method by monitoring system suitability and/or the analysis of samples. It is part of the formal methods validation process.

Ruggedness. A dated term, now commonly accepted as intermediate precision. See definition of **intermediate precision.**

Selectivity. The ability of the method to separate the analyte from other components that may be present in the sample, including impurities. Determination of selectivity normally includes analyzing placebo, blank, media for dissolution, dilution solvent, and mobile phase injections. Also, no chromatographic peaks, such as related compounds, should interfere with the analyte peak or internal standard peak, if applicable. Selectivity=separate and show every component in the sample.

Signal to Noise (S/N). In chromatography the measure of average baseline noise (e.g., peak-to-peak) to the signal given by an analyte peak. S/N calculations are performed when determining the LOD and LOQ.

Specification. The quality standards (e.g., tests, analytical procedures, and acceptance criteria) provided in an approved application to confirm the quality of drug substances, drug products, intermediates, raw materials, reagents, and other components including in-process materials.

Spiked placebo. Preparation of a sample to which known quantities of analyte are added to placebo material. Performed during validation to generate accurate and reproducible samples used to demonstrate recovery from the sample matrix.

Spiking. The addition of know amounts of a known compound to a standard, sample, or placebo, typically for the purpose of confirming the performance of an analytical procedure or the calibration of an instrument.

Specificity. The ability to assess unequivocally the analyte in the presence of components that may be expected to be present such as impurities, degradation products, and excipients. There must be inarguable data for a method to be specific. Specific=measure only the desired component without interference from other species that might be present; separation is not necessarily required.

Stability indicating methodology. A validated quantitative analytical procedure or set of procedures that can detect the changes with time in the pertinent properties (e.g., active ingredient, preservative level, or appearance of degradation products) of the drug substance and drug product

Stability indicating assay. An assay that accurately measures the component of interest [the active ingredient(s) or degradation products] without interference form other degradation products, process impurities, excipients, or other potential interfering substances.

Stability indication profile. A set of procedures or assays that collectively detect changes with time, although may not do so individually.

Standard and sample solution stability. Established under normal benchtop conditions, normal storage conditions, and sometimes in the instrument (e.g., an HPLC autosampler) to determine if special storage conditions are necessary, for instance, refrigeration or protection from light. Stability is determined by comparing the response and impurity profile from aged standards or samples to that of a freshly prepared standard and to its own response from earlier time points. Note that these are short-term studies and are not intended to be part of the stability indication assessment or product stability program.

Stressed studies. See **Forced degradation studies**.

System suitability. Evaluation of the components of an analytical system to show that the performance of a system meets the standards required by a method. A system suitability evaluation usually contains its own set of parameters. For chromatographic assays, these may include tailing factors, resolution, and precision of standard peak areas, and comparison to a confirmation standard, capacity factors, retention times, theoretical plates, and calibration curve linearity.

Tailing factor. A measure of peak asymmetry. Peaks with a tailing factor of >2 are usually considered to be unacceptable due to difficulties in determine peak start and stop points which complicates integration. Tailing peaks are an indication that the chromatographic conditions for a separation have not been properly optimized.

Test method. An approved, detailed procedure describing how to test a sample for a specified attribute (e.g., assay), including the amount required, instrumentation, reagents, sample preparation steps, data generation steps and calculations use for evaluation.

Theoretical plates. A dimensionless quantity used to express the efficiency or performance of a column under specific conditions. A decrease in theoretical plates can be an indication of HPLC column deterioration.

Transcription accuracy verification (TAV). The process where the transcription of data from one location to another is confirmed by a second party. Important with methods validation reports which summary results, which, are often transcribed and not linked to raw data.

USP <1225> Category I. One of four method categories for which validation data should be required. Category I methods include analytical methods for quantitation of major components of bulk drug substances or active ingredients (including preservatives) in finished pharmaceutical products. Category I methods are typically referred to as assays.

USP <1225> Category II. One of four method categories for which validation data should be required. Category II methods include analytical methods for determination of **impurities** in bulk drug substances or **degradation compounds** in finished pharmaceutical products. These methods include quantitative assays and limits tests.

USP <1225> Category III. One of four method categories for which validation data should be required. Category III methods include analytical methods for determination of performance characteristics (e.g., **dissolution**, drug release).

USP <1225> Category IV. One of four method categories for which validation data should be required. Category IV methods include analytical methods used as *identification tests*.

Validation characteristics. See **Analytical performance characteristics.**

Validation parameters. See **Analytical performance characteristics.**

Validation report. A summary of experiments and results that demonstrate the method is suitable for its intended use, approved by responsible parties.

ATTACHMENT II: USP METHOD CATEGORIES AND ANALYTICAL PERFORMANCE CHARACTERISTICS EVALUATION AND TESTING REQUIRED FOR VALIDATION

A USP method category refers to the most common types of analyses (types of methods) for which validation data should be required.

Analytical performance characteristics refers to those characteristics of an analytical method which define its performance as an analytical technique. The USP identifies these performance characteristics as (1) accuracy, (2) precision, (3) specificity, (4) detection limit, (5) quantitation limit, (6) linearity, and (7) range, and indicates that these must be evaluated during methods validation. Moreover, additional validation testing is frequently required and most often includes (8) robustness testing, (9) system suitability determination, (10) forced degradation studies, (11) solution stability testing, (12) filter retention studies, and (13) extraction efficiency studies. Definitions of each term and how they are determined during validation are presented in Table 1.

TABLE 1

Analytical Performance Characteristic or Validation Test	Definition	Determination During Validation
1. Accuracy (recovery)	Expresses the closeness of agreement between the value found and the value that is accepted as either a conventional true value or an accepted reference value. It may often be expressed as the recovery by the assay of known, added amounts of analyte.	Samples (spiked placebos) are prepared normally covering 50% to 150% of the nominal sample preparation concentration. These samples are analyzed and the recoveries of each are calculated.
2. Precision (2 parts):		
Repeatability (method precision)	The variation experienced by a single analyst on a single instrument. Repeatability does not distinguish between variation from the instrument or system alone and from the sample preparation process.	Multiple replicates of an assay composite sample are analyzed using the analytical method. The recovery value is calculated and reported for each value.

TABLE 1   *(Continued)*

Analytical Performance Characteristic or Validation Test	Definition	Determination During Validation
Intermediate precision	Refers to variations within a laboratory as with different days, with different instruments, by different analysts, and so forth. Formally known as ruggedness.	A second analyst repeats the repeatability analysis on a different day using different conditions and different instruments. The recovery values are calculated and reported. A statistical comparison is made to the first analyst's results.
3. Specificity	Specificity is the ability to assess unequivocally the analyte in the presence of components that may be expected to be present such as impurities, degradation products, and excipients. There must be inarguable data for a method to be specific. Specificity measures only the desired component without interference from other species that might be present; separation is not necessarily required.	Analyze blanks, sample matrix (placebo), and known related impurities to determine whether interferences occur. Also, demonstrated during forced degradation studies. Relative response factors are also determined.
4. Detection limit	The detection limit (DL) or Limit of detection (LOD) of an individual procedure is the lowest amount of analyte in a sample that can be detected but not necessarily quantitated as an exact value. The LOD is a parameter of limit test (tests that only determine if the analyte concentration is above or below a specification limit).	In analytical procedures that exhibit baseline noise, the LOD can be based on a signal-to-noise ratio (3:1), which is usually expressed as the concentration (e.g., percentage, parts per billion) of analyte in the sample. There are several ways in which it can be determined, but usually it involves injecting samples which generate S/N of 3:1 and estimating the DL.
5. Quantitation limit	The quantitation limit (QL) or limit of quantitation (LOQ) of an individual analytical procedure is the lowest amount of analyte in a sample	For analytical procedures that exhibit baseline noise the LOQ is generally estimated from a determination of signal-to-noise ratio (10:1) and is usually

(Continued)

TABLE 1   *(Continued)*

Analytical Performance Characteristic or Validation Test	Definition	Determination During Validation
	that can be quantitatively determined with suitable precision and accuracy. The quantitation limit is a parameter of quantitative assays for low concentrations of compounds in sample matrices and is used particularly for the determination of impurities and/or degradation products. It is usually expressed as the concentration (e.g., percentage, parts per million) of analyte in the sample.	confirmed by injecting standards which give this S/N ratio and have acceptable % RSDs as well.
6. Linearity	Evaluates the analytical procedure's ability (within a given range) to obtain a response that is directly proportional to the concentration (amount) of analyte in the sample. If the method is linear, the test results are directly, or by well-defined mathematical transformation, proportional to the concentration of analyte in samples within a given range. Linearity is usually expressed as the confidence limit around the slope of the regression line. The line is calculated according to an established mathematical relationship from the test response obtained by the analysis of samples with varying concentrations of analyte. Note that this is different from *Range* (sometimes referred to as *linearity of method*), which is evaluated using samples and must encompass the specification range of the component assayed in the drug product.	Linearity may be established for all active substances, preservatives, and expected impurities. Evaluation is performed on standards.

TABLE 1   *(Continued)*

Analytical Performance Characteristic or Validation Test	Definition	Determination During Validation
7. Range	The interval between the upper and lower concentrations (amounts) of analyte in the sample (including these concentrations) for which it has been demonstrated that the analytical procedure has a suitable level of precision, accuracy, and linearity. Range is normally expressed in the same units as test results (e.g., percent, parts per million) obtained by the analytical method.	Range (sometimes referred to as *linearity of method*) is evaluated using samples (usually spiked placebos) and must encompass the specification range of the component assayed in the drug product.
8. Robustness testing	The measure of the ability of an analytical method to remain unaffected by small but deliberate variations in method parameters (e.g., pH, mobile phase composition, temperature, instrument settings) and provides an indication of its reliability during normal usage.	Robustness testing is a systematic process of varying a parameter and measuring the effect on the method by monitoring system suitability and/or the analysis of samples. It is part of the formal method validation process.
9. System suitability determination	System suitability is the evaluation of the components of an analytical system to show that the performance of a system meets the standards required by a method. A system suitability evaluation usually contains its own set of parameters; for chromatographic assays, these may include tailing factors, resolution, and precision of standard peak areas, and comparison to a confirmation standard, capacity factors, retention times, theoretical plates, and calibration curve linearity.	Where applicable, system suitability parameters are calculated, recorded, and trended throughout the course of the validation. Final values are then determined from this history.
10. Forced degradation studies	Studies undertaken to degrade the sample (e.g., drug product or API) deliberately. These	Samples or drug product (spiked placebos) and drug substance are exposed to heat, light, acid,

(Continued)

TABLE 1 *(Continued)*

Analytical Performance Characteristic or Validation Test	Definition	Determination During Validation
	studies are used to evaluate an analytical method's ability to measure an active ingredient and its degradation products without interference.	base, and oxidizing agent to produce 10%–30% degradation of the active. The degraded samples are then analyzed using the method to determine if there are interferences with the active or related compound peaks.
11. Solution stability studies	The stability of standards and samples is established under normal benchtop conditions, normal storage conditions, and sometimes in the instrument (e.g., an HPLC autosampler) to determine if special storage conditions are necessary, for instance, refrigeration or protection from light.	Stability is determined by comparing the response and impurity profile from aged standards or samples to that of a freshly prepared standard and to its own response from earlier time points.
12. Filter retention studies	A comparison of filtered to unfiltered solutions during a methods validation to determine whether the filter being using retains any active compounds or contributes unknown compounds to the analysis.	Blank, sample, and standard solutions are analyzed with and without filtration. Comparisons are made in recovery and appearance of chromatograms.
13. Extraction efficiency studies	Extraction efficiency is the measure of the effectiveness of extraction of the drug substance from the sample matrix. Studies are conducted during method validation to determine that the sample preparation scheme is sufficient to ensure complete extraction without being unnecessarily excessive.	Extraction efficiency is normally investigated by varying the shaking or sonication times (and/or temperature) as appropriate during sample preparation on manufactured (actual) drug product dosage forms.

The definition for each USP category and corresponding analytical performance characteristics and additional validation testing requirements are show in the following list.

1. CATEGORY I METHODS

1.1. *Definition of USP <1225> Category I Methods.* Category I methods include analytical methods for quantitation of major components of bulk drug substances or active ingredients (including preservatives) in finished pharmaceutical products. Category I methods are typically referred to as assays.

1.2. *Analytical Performance Characteristics and Additional Validation Testing Required for Category I Methods.* The performance characteristics and additional validation testing evaluated for Category I methods during validation include:

    1.2.1. Accuracy (recovery)

    1.2.2. Precision (2 parts):

        1.2.2.1.1. Repeatability (method precision)

        1.2.2.1.2. Intermediate precision

    1.2.3. Specificity

    1.2.4. Detection limit—NOT REQUIRED

    1.2.5. Quantitation limit—NOT REQUIRED

    1.2.6. Linearity

    1.2.7. Range

    1.2.8. Robustness

    1.2.9. System suitability determination

    1.2.10. Forced degradation studies

    1.2.11. Solution stability studies

    1.2.12. Filter retention studies

    1.2.13. Extraction efficiency studies

2. CATEGORY II METHODS

2.1. *Definition USP <1225> Category II.* Category II methods include analytical methods for determination of impurities in bulk drug substances or degradation compounds in finished pharmaceutical products. These methods include quantitative assays and limits tests.

2.2. *Analytical Performance Characteristics and Additional Validation Testing Required for Category II Methods–Quantitative Tests.* The performance characteristics and additional validation testing evaluated for Category II methods—quantitative assays during validation include:

    2.2.1. Accuracy (recovery)

    2.2.2. Precision (2 parts):

        2.2.2.1.1. Repeatability (method precision)

        2.2.2.1.2. Intermediate precision

2.2.3. Specificity

2.2.4. Detection limit

2.2.5. Quantitation limit

2.2.6. Linearity

2.2.7. Range

2.2.8. Robustness

2.2.9. System suitability determination

2.2.10. Forced degradation studies

2.2.11. Solution stability studies

2.2.12. Filter retention studies

2.2.13. Extraction efficiency studies

2.3. *Analytical Performance Characteristics and Additional Validation Testing Required for Category II Methods–Limits Tests*. The performance characteristics and additional validation testing evaluated for Category II methods–limit tests during validation include:

2.3.1. Accuracy (recovery)–MAY BE REQUIRED

2.3.2. Precision (2 parts):

2.3.2.1.1. Repeatability (method precision)—NOT REQUIRED

2.3.2.1.2. Intermediate precision—NOT REQUIRED

2.3.3 Specificity

2.3.4. Detection limit

2.3.5. Quantitation limit—NOT REQUIRED

2.3.6. Linearity—NOT REQUIRED

2.3.7. Range—MAY BE REQUIRED

2.3.8. Robustness—NOT REQUIRED

2.3.9. System suitability determination

2.3.10. Forced degradation studies—NOT REQUIRED

2.3.11. Solution stability studies

2.3.12. Filter retention studies

2.3.13. Extraction efficiency studies—MAY BE REQUIRED

3. CATEGORY III METHODS

3.1. *Definition USP <1225> Category III.* Category III methods include analytical methods for determination of performance characteristics (e.g., **dissolution**, drug release).

3.2. *Analytical Performance Characteristics and Additional Validation Testing Required for Category III Methods*. The performance characteristics and additional validation testing evaluated for Category III methods during validation include:

3.2.1. Accuracy (recovery)—MAY BE REQUIRED

3.2.2. Precision (2 parts):

3.2.2.1.1. Repeatability (method precision)

3.2.2.1.2. Intermediate precision

3.2.3. Specificity—MAY BE REQUIRED

3.2.4. Detection limit—MAY BE REQUIRED

3.2.5. Quantitation limit—MAY BE REQUIRED

3.2.6. Linearity—MAY BE REQUIRED

3.2.7. Range—MAY BE REQUIRED

3.2.8. Robustness—MAY BE REQUIRED

3.2.9. System suitability determination

3.2.10. Forced degradation studies—NOT REQUIRED

3.2.11. Solution stability studies

3.2.12. Filter retention studies

3.2.13. Extraction efficiency studies—NOT REQUIRED

4. CATEGORY IV METHODS

4.1. *Definition USP <1225> Category IV.* Category IV methods include analytical methods used as **identification tests**.

4.2. *Analytical Performance Characteristics and Additional Validation Testing Required for Category IV Methods.* The performance characteristics and additional validation testing evaluated for Category IV methods during validation include:

4.2.1. Accuracy (recovery)—NOT REQUIRED

4.2.2. Precision (2 parts):

4.2.2.1.1. Repeatability (method precision)—NOT REQUIRED

4.2.2.1.2. Intermediate precision—NOT REQUIRED

4.2.3. Specificity

4.2.4. Detection limit—NOT REQUIRED

4.2.5. Quantitation limit—NOT REQUIRED

4.2.6. Linearity—NOT REQUIRED

4.2.7. Range—NOT REQUIRED

4.2.8. Robustness—NOT REQUIRED

4.2.9. System suitability determination

4.2.10. Forced degradation studies—NOT REQUIRED

4.2.11. Solution stability studies

4.2.12. Filter retention studies

4.2.13. Extraction efficiency studies—MAY BE REQUIRED

Table 2 correlates method category to which analytical performance characteristics are required to be determined during validati

TABLE 2

Analytical Performance Characteristic or Validation Test	USP Method Category I (Assay)	USP Method Category II Quantitative Tests (Related Compounds)	USP Method Category II Limit Tests (Related Compounds)	USP Method Category III Dissolution	USP Method Category IV (ID Tests)
1. Accuracy (recovery)	Yes	Yes	*	*	No
2. Precision (2 parts)					
a. Method precision	Yes	Yes	No	Yes	No
b. Intermediate precision	Yes	Yes	No	Yes	No
3. Specificity	Yes	Yes	Yes	*	Yes
4. Detection limit	No	Yes	Yes	*	No
5. Quantitation limit	No	Yes	No	*	No
6. Linearity	Yes	Yes	No	*	No
7. Range	Yes	Yes	*	*	No
8. Robustness testing	Yes	Yes	No	*	No
9. System suitability determination	Yes	Yes	Yes	Yes	Yes
10. Forced degradation studies	Yes	Yes	No	No	No
11. Solution stability studies	Yes	Yes	Yes	Yes	Yes
12. Filter retention studies	Yes	Yes	Yes	Yes	Yes
13. Extraction efficiency studies	Yes	Yes	*	No	*

* If applicable.

ATTACHMENT III: RECOMMENDED EXPERIMENTAL PROCEDURES AND ACCEPTANCE CRITERIA FOR CATEGORY I METHODS (ASSAY)

Category I Methods (Assay)

Analytical Performance Characteristic or Validation Test	Recommended Procedure	Recommended Acceptance Criteria
1. Accuracy (Recovery)	**If not using the samples and results from the range experiments perform the following:**	The percent recovery of the spiked placebos should be within $100 \pm 2.0\%$ for the average of each set of three weights.
	Prepare 15 samples by weighing an appropriate amount of placebo (the placebo remains 100% of method concentration in all samples), with respect to the concentration specified in the method being validated.	Each individual sample recovery should lie within the range of 98%–102%.
	Spike the placebo samples with one stock active solution (using diluent as per the sample preparation) at each of the five levels.	Coefficient of determination (r^2) should be greater than 0.9998
	Prepare three (3) replicate weights for each level.	There should be no curvature in the residuals plot.
	Prepare three samples at about 50%.	The y intercept should not significantly depart from zero (e.g., area response of y intercept should be less than 5% of the response of the nominal 100% concentration value.)
	Prepare three samples at about 70%.	
	Prepare three samples at about 100%.	
	Prepare three samples at about 130%.	
	Prepare three samples at about 150%.	

(Continued)

97

Category I Methods (Assay) *(Continued)*

Analytical Performance Characteristic or Validation Test	Recommended Procedure	Recommended Acceptance Criteria
	Inject each sample three times and analyze according to the analytical method, adequately bracketed by standard.	
	Inject samples from the lowest concentration to the highest concentration.	
	Calculate the % RSD for each individual weight at each level.	
	Plot the analyte concentration for each individual weight versus the signal response (average of each set of injections).	
	Perform linear regression analysis, but do not include the origin as a point made and do not force the line through the origin.	
	Plot the sign and magnitude of the residuals versus analyte concentration.	
	Check residual plot for outlying values and curvature.	
	Evaluate y intercept to determine if there is significant departure from zero.	
	Calculate the recovery for each individual sample weight (average of the three injections).	
	Calculate the average recovery of the three sample weights at each concentration level.	

Notes:

Separate recovery studies should be performed for each formulation type of various dosage forms (i.e., tablets/capsules, oral/sublingual, etc.).

2. Precision (2 parts):

Repeatability (method precision)	Prepare six replicate samples solutions from the same assay composite sample according to the analytical method.	The % RSD of the assay or recovery values should not be greater than 2.0%.
	Analyze the samples according to the analytical method and make two injections of each sample.	
	Calculate the assay results (% recovery) for each sample.	
	Calculate the mean and relative standard deviations (% RSD) of the six sample preparations.	
Intermediate precision	**With the following restrictions, have a second analyst execute the following steps:**	The % RSD of the assay/recovery values generated by a single analyst should not be greater than 2.0%.
	Perform the work on different days	The % RSD of the combined assay/recovery values generated by both analysts over both days should not be greater than 3.0%.
	Perform the work using different operating conditions and different instruments when possible (e.g., column, apparatus, reagents).	
	If possible use a different manufacturer's instrument.	
	Prepare six replicate samples solutions from the same assay composite sample according to the analytical method.	
	Analyze the samples according to the analytical method; make two injections of each sample.	

(Continued)

Category I Methods (Assay) (Continued)

Analytical Performance Characteristic or Validation Test	Recommended Procedure	Recommended Acceptance Criteria
	Calculate the assay results (% recovery) for each sample.	
	Calculate the mean and relative standard deviations (% RSD) of the six sample preparations.	
	Calculate the average assay/recovery values and the % RSD of all the individual sample preparations generated by both analysts over both days.	
	Notes: Precision should be investigated with homogenous samples.	
	If it was not possible to obtain homogenous samples, artificially prepared samples or samples solutions can be used.	
	If the method is used for a range of dosage types (tablets/capsules, oral/sublingual, etc.), the method should be validated for method precision with each dosage type.	
	The same grind method should used by both analysts if a grind was necessary for sample preparation (e.g., tablets).	
	If there are multiple potencies, the precision method should be performed on the low and high potency.	

100

If there is only one strength of a product, the method precision method should be performed on two different lots of that one strength.

A chromatogram of a typical standard, a typical system suitability standard, and a typical sample should be included in the validation report.

3. Specificity

Make an injection of a blank/diluent during each chromatographic run to ensure there are no interferences.

If an internal standard was used, inject the internal standard by itself to confirm specificity.

Prepare a placebo for each potency. If the formulations are dose proportional, placebo interference may be performed on only one potency.

Inject each placebo preparation twice.

Confirm that no peaks can be attributed to the blank/diluent or placebo.

Define any peaks observed by RRT indexed to the active component.

Prepare and inject twice, samples of each individual impurity at the impurity/degradant (e.g., related compound) specification limit.

Prepare two separate spiked solutions containing the active at 100% and each impurity (from two separate impurity preparations) at the limit.

Inject the spiked samples twice to confirm specificity.

Confirmed that no interference in the elution zone of the active occurred from the blank/diluent, internal standard (if applicable), or the placebo.

Confirm that the impurities/degradants do not elute in the elution zone of the active.

(Continued)

Category I Methods (Assay) (Continued)

Analytical Performance Characteristic or Validation Test	Recommended Procedure	Recommended Acceptance Criteria
	Calculate the relative response factor (RRF) for each impurity using the spiked solution preparation.	
4. Detection limit	Not performed for category I Methods (Assay).	
5. Quantitation limit	Not performed for Category I Methods (Assay).	
6. Linearity	Prepare five standard solutions of the analyte at ~50%, 70%, 100%, 130%, and 150% of the method concentration using serial dilutions from a stock solution.	Coefficient of determination (r^2) should be greater than 0.9999.
	Make three injections at each concentration, adequately bracketed by the standard.	There should be no curvature in the residuals plot.
	Make sure to inject samples from the lowest concentration to the highest concentration to reduce the effects, if any, of carryover from the higher concentration samples.	The y intercept should not significantly depart from zero (e.g., area response of y intercept should be less than 5% of the response of the midrange concentration value).
	Calculate the % RSD at each concentration.	
	Plot the analyte concentration for each set of dilutions separately versus the signal response (average of each set of injections).	
	Perform linear regression analysis, but do not include the origin as a point made and do not force the line through the origin.	

Plot the sign and magnitude of the residuals versus analyte concentration.

Check residual plot for outlying values and curvature.

Evaluate y intercept to determine if there is significant departure from zero.

Notes:

If method is to be used for multiple analyte concentrations, ensure linearity was examined from 50% of the lowest nominal concentration to 150% of highest nominal concentration.

Coefficient of determination (r^2) should be greater than 0.9998.

There should be no curvature in the residuals plot.

The y intercept should not significantly depart from zero (e.g., area response of y intercept should be less than 5% of the response of the nominal 100% concentration value).

7. Range

If not using the samples and results from the accuracy (recovery) experiments perform the following:

Prepare 15 samples by weighing an appropriate amount of placebo (the placebo remains 100% of method concentration in all samples), with respect to the concentration specified in the method being validated.

Spike the placebo samples with one stock active solution (using diluent as per the sample preparation) at each of the five levels.

Prepare three (3) replicate weights for each level.

Prepare three samples at about 50%.

Prepare three samples at about 70%.

Prepare three samples at about 100%.

(Continued)

Category I Methods (Assay) (*Continued*)

Analytical Performance Characteristic or Validation Test	Recommended Procedure	Recommended Acceptance Criteria
	Prepare three samples at about 130%.	
	Prepare three samples at about 150%.	
	Inject each sample three times and analyze according to the analytical method, adequately bracketed by standard.	
	Inject samples from the lowest concentration to the highest concentration.	
	Plot the analyte concentration for each individual weight versus the signal response (average of each set of injections).	
	Calculate the % RSD for each individual weight at each level.	
	Perform linear regression analysis, but do not include the origin as a point and do not force the line through the origin.	
	Plot the sign and magnitude of the residuals versus analyte concentration.	
	Check residual plot for outlying values and curvature.	
	Evaluate y intercept to determine if there is significant departure from zero.	

Notes:

If there are multiple potencies, perform range experiments for both the lowest and highest potency.

If the potencies are dose proportional, then the linearity may have been performed using one potency.

If compounds are of low solubility, prepare dry spiked placebos or dissolve the compound in a small amount of solvent in which it is soluble.

The range samples may also be used for accuracy/recovery.

8. Robustness testing	**Mobile phase variation**	Measure appropriate figures of merit (e.g., resolution, tailing factor, theoretical plates, and capacity factor) for each variation experiment.
	Using the mobile phase specified in the test method:	Determine the suitability of the method under each modification determined by taking into account peak shape, retention time, system pressure, and system suitability parameters.
	Increase each [major] component by 5%, inject the system suitability standard twice, and measure the appropriate figures of merit.	Determine which system suitability parameters are important to the overall function of the method.
	Decrease each [major] component by 5%, inject the system suitability standard twice, and measure the appropriate figures of merit.	
	Increase each [major] component by 10%, inject the system suitability standard twice, and measure the appropriate figures of merit.	Establish limits for critical parameters.
	Decrease each [major] component by 10%, inject the system suitability standard twice, and measure the appropriate figures of merit.	For column variability ensure all columns used in validation are commercially available.
	Increase each [minor] component by 15%, inject the system suitability standard twice, and measure the appropriate figures of merit. (*Note:* Minor components are those with less than 10 mL/L.)	

(Continued)

Category I Methods (Assay) (Continued)

Analytical Performance Characteristic or Validation Test	Recommended Procedure	Recommended Acceptance Criteria
	Decrease each [minor] component by 15%, inject the system suitability standard twice, and measure the appropriate figures of merit.	Ensure that three columns from at least two different lots of packing material obtained and used.
	Increase each [minor] component by 30%, inject the system suitability standard twice, and measure the appropriate figures of merit.	Ensure that a brand new column and an old column (>500 injections) are obtained and used.
	Decrease each [minor] component by 30%, inject the system suitability standard twice, and measure the appropriate figures of merit.	Ensure the retention times are similar on each of the three columns.
	HPLC column temperature variation	
	Using a column heater:	
	Inject the system suitability standard injected twice at the temperature stated in the method, and measure the appropriate figures of merit.	
	Inject the system suitability standard injected twice at +5° above stated method temperature and measure the appropriate figures of merit.	
	Inject the system suitability standard injected twice at −5° above stated method temperature and measure the appropriate figures of merit.	
	Mobile phase flow-rate variation	
	Inject the system suitability standard twice at a 10% increase in flow rate, and measure the appropriate figures of merit.	

Inject the system suitability standard twice at a 10% decrease in flow rate, and measure the appropriate figures of merit.

Inject the system suitability standard injected twice at a 25% increase in flow rate, and measure the appropriate figures of merit.

Inject the system suitability standard injected twice at a 25% decrease in flow rate, and measure the appropriate figures of merit.

For buffer pH variation

Increase the mobile phase 0.25 pH units, inject system suitability standard injected twice, and measure the appropriate figures of merit.

Decrease the mobile phase 0.25 pH units, inject system suitability standard injected twice, and measure the appropriate figures of merit.

HPLC column variation

Using three columns from at least two different lots of packing material, inject the system suitability standard injected twice on each column, and measure the appropriate figures of merit.

Using a brand new column and an old column (> 500 injections), inject system suitability standard injected twice on each column, and measure the appropriate figures of merit.

9. System suitability determination

Where applicable, calculate system suitability (capacity factor, tailing factor, resolution, theoretical plates, and reproducibility) for each chromatographic run during methods validation.

The following minimum criteria are suggested:

$k' \geq 2.0$, where k' is capacity factor

(Continued)

107

Category I Methods (Assay)   (Continued)

Analytical Performance Characteristic or Validation Test	Recommended Procedure	Recommended Acceptance Criteria
	Identify upper and lower limits for each parameter by analyzing values obtained throughout the validation.	$T \leq 2$, where T is tailing factor
		$R > 1.5$, where R is resolution
	Establish minimum criteria from this range.	$N \geq 1000$ plates, where N is theoretical plates per column
		% RSD $\leq$ 2.0%, where % RSD is the % RSD.

10. Forced degradation studies — **The following degradation conditions are recommended as a starting point:**

Degradation Reaction	Typical Conditions	Recommended Acceptance Criteria
Acid hydrolysis	Sample in aqueous acid or acidified solvent at ~0.5 N up to 24 hours (or)	Confirm that peak purity of the degraded spiked-placebo sample main peak was performed using UV or MS analysis.
	Heat/reflux or UV radiation in ~0.5 N HCl up to 24 hours.	Ensure peak purity was performed for each degradation condition and that the samples have degraded sufficiently or degradation had stopped.
Base hydrolysis	Sample in aqueous base or basic solvent at ~0.5 N up to 24 hours (or)	Evaluate if spectral overlay of the sufficiently degraded impurities/degradants spiked placebos, was used in addition to peak purity to demonstrate that the degradants are resolved from the analyte.
	Heat/reflux or UV radiation in ~0.5 N NaOH up to 24 hours.	
Oxidation	Treat with ~3% H_2O_2 up to 24 hours (or)	Determine if relative retention times of the degradants were calculated.
	UV irradiation in ~3% H_2O_2 up to 24 hours.	
Light decomposition (photolysis)	Expose to high-intensity UV light in suitable increments, up to 24 hours.	
Thermal decomposition (pyrolysis)	Expose to ~100° C heat in suitable increments, up to 24 hours.	

Obtain or prepare solutions of ~0.5 N HCL, ~0.5 N NaOH, ~3% H_2O_2, and purified water.

Prepare blank samples for each condition, including light and heat.

Prepare a standard, spiked placebo and unspiked placebo as appropriate in preparation for degradation.

Analyze samples and blanks according to the method outlined in the validation prior to degradation, establishing an initial time point.

Expose these standards, spiked placebos, and unspiked placebos to acid, base, oxidation, heat, and light using the condition guidelines listed above.

Neutralize acid and base samples prior to analysis by pipetting in an amount of acid or base solution equal to the sample aliquot, and then diluting to volume with water, diluent, or mobile phase.

Analyze samples and blanks at varying intervals over 24 hours, according to the method outlined in the validation following degradation. Assess whether sufficient degradation (10%–30%) has occurred.

Calculate the percent recovery of the appropriate solutions to determine the extent of degradation.

If the acid, base, and peroxide solutions were not sufficiently degraded in 24 hours, the acid, base, and peroxide solutions should have been exposed to heat/light, until at least 10%–30% degradation is achieved or 24 hours of exposure has elapsed. Moreover, if the samples are overdegraded, lessening the length and severity of the conditions to obtain the 10%–30% is acceptable.

Assess if the chromatograms of the sufficiently degraded spiked placebo overlaid with the degraded placebo under each degradation condition are included with the validation report.

Assess whether analysis of expired finished product was undertaken to confirm specificity of the method (if necessary).

(Continued)

109

Analytical Performance Characteristic or Validation Test	Recommended Procedure	Recommended Acceptance Criteria
	Perform peak purity analysis on the main analyte peak using diode array and/or mass spectra analysis.	
	Capture and display chromatograms for each degradation on appropriately degraded (10%–30%) samples.	
	Capture and display peak purity analyses for each degradation condition.	
	Determine whether the method is specific for the degraded samples.	
	Notes: Forced degradation studies are designed to produce potential degradation products which may be encountered in real-world scenarios. Degradants generated may or may not be what is seen during stability studies.	
	Performing the actual degradation in these studies is not an exact science and may require modifying conditions to obtain the desired 10%–30% degradation. However, if the maximum conditions listed here do not produce degradation, then it is not necessary to continue the experiments until degradation occurs. A statement is then added to the report confirming the stable nature of the molecules.	

For multicomponent formulations (e.g., with more than one active), individual active solutions should be made for each component. For product families that utilize the same excipients, forced degradation should be formed on only one formulation.

If dyes are utilized in the product information, placebos for each different formulation can be prepared and used for the forced degradation study. At the discretion of the front-line supervisor, if dyes are utilized in the product formulation, a worst case placebo can be used for the forced degradation study.

Acid and base samples should be prepared at 2X times the nominal concentration, so that they may be neutralized prior to analysis.

If the analyte is not soluble in aqueous solutions, a small amount of suitable organic solvent may be used to dissolve the sample.

The acid, base, and oxidizing solutions may be made up in solvent.

The concentration of the acid, base, and peroxide solutions may be reduced or the sample concentration may be lowered with subsequent increase of the injection volume to maintain the appropriate amount of material on column.

If deemed necessary by the supervisor, forced degradation studies may also be performed on dry powder samples.

(Continued)

Category I Methods (Assay) (*Continued*)

Analytical Performance Characteristic or Validation Test	Recommended Procedure	Recommended Acceptance Criteria
	If forced degradation demonstrates lack of specificity, analysis of expired finished product may be used to prove that the forced degradation conditions are not producing real degradant peaks.	
	Targeted recovery for the initial samples prior to degradation should be in the 95%–105% range.	
	The run time for forced degradation samples should be sufficiently long to observe the retention time of the latest eluting active or degradant.	
	For product families that utilize the same excipient, forced degradation should be formed on only one formulation.	
11. Solution stability studies	Prepare fresh standard and spiked placebo (or actual tablet/capsule) as per the test method. Ensure that both stock and working solutions are available for analysis.	For assay level standards, the fresh standard and the verification standard should not differ more that 1.0%.
	Analyze these solutions analyzed as per the test method, establishing a time zero value for each.	For the assay level, the standard and sample solutions are considered sufficiently stable over time if the recovery value does not vary more than $\pm 1.5\%$ (absolute) from the initial result.
	Place an aliquot of each solution in clear glassware and exposed to ambient (benchtop) conditions.	
	Place an aliquot of each solution placed in amber glassware and exposed to ambient (benchtop) conditions.	

112

Place an aliquot of each solution in clear glassware and place in a refrigerator.

Analyze these samples versus fresh standard every 24 hours for at least two (2) days.

Make two injections made of each solution.

Calculate the percent recoveries calculated for all solutions.

Notes:
After two days, samples may be evaluated at intervals at the analyst's discretion.

A fresh standard and fresh verification standard should be prepared each day.

12. **Filter retention studies**

Prepare three placebos for the lowest potency, spiked with dry active at 100% of the nominal assay concentration.

For the spiked placebos, the percentage recovery should be $100 \pm 2.0\%$ for the average of each set of three weights.

Where necessary, three product sample preparations (tablets, capsules, drug substance, etc.) may be substituted for the dry spiked placebos.

For each individual spiked placebo, the recovery should be 98%–102%.

Filter portions of each individual spiked placebo sample through at least two (2) prospective filters. Ensure the filters are commercially available. If this is a revalidation, one of the prospective filters one that is specified in the respective monograph or test method.

For an acceptable filter, the difference between the filtered sample and the centrifuged sample NMT 1.5% (absolute).

(Continued)

Category I Methods (Assay) *(Continued)*

Analytical Performance Characteristic or Validation Test	Recommended Procedure	Recommended Acceptance Criteria
	Centrifuge an aliquot of each individual spiked placebo (e.g., 3000 rpm or 10 minutes suggested)	
	Inject all three samples three times each and analyze according to the analytical method adequately bracketed by the standard.	
	Calculate the recovery of each individual sample weight and % RSD of the replicate injections.	
	Calculate the average recovery for each filter.	
	Calculate the percent difference of the average result for each filter versus the average result of the centrifuged sample.	
13. Extraction efficiency studies	Using aged samples, preferably an expired lot, obtain 12 sample weights.	Establish an extraction range for the sample preparation procedure after evaluation of the data.
	Prepare the first set of three samples prepared as per the sample procedure in the method.	
	Prepare the second set of three samples prepared and extracted longer than the first set.	
	Prepare the third set of three samples prepared and extracted longer than the second set.	

Prepare the fourth set of three samples prepared and extracted for a shorter period of time than that specified extraction time in the procedure.

Calculate the results in percent recovered.

Notes:

It should be demonstrated that the extraction of the drug from the sample matrix is sufficient to ensure complete extraction without being unnecessarily excessive. This is investigated by varying the shaking or sonication times (and/or temperature) as appropriate.

If the assay is a grind method and the content uniformity method is a drop method, one unit from the low potency and the high potency should be analyzed and the extraction process should be monitored to document at what point the tablet disintegrated.

ATTACHMENT IV: RECOMMENDED EXPERIMENTAL PROCEDURES AND ACCEPTANCE CRITERIA FOR CATEGORY II METHODS (IMPURITY AND DEGRADANTS)

Category II (Impurity and Degradants) Methods

Analytical Performance Characteristic or Validation Test	Recommended Procedure	Recommended Acceptance Criteria
1. Accuracy (recovery)	**If not using the samples and results from the range experiments perform the following:**	
	Prepare 15 samples by weighing an appropriate amount of placebo (the placebo remains 100% of method concentration in all samples) with respect to the concentration specified in the method being validated.	The individual sample recovery of each impurity should be within 75% and 125%
	Spike each of the 15 placebo samples with active solution at 100% of the nominal active concentration.	The average percent recovery for each impurity should lie within the range of 75% to 125%.
	Prepare spiking solutions for each of the impurities.	Coefficient of determination (r^2) should be greater than 0.995
	Spike the 15 placebo samples with each impurity in the following fashion:	There should be no curvature in the residuals plot.
	• Prepare three replicate samples at about the LOQ concentration.	The y intercept should not significantly depart from zero (e.g., area response of y intercept should be less than 5% of the response of the nominal 100% concentration value).
	• Prepare three replicate samples at about 25% of the impurity/degradant limit for each impurity.	
	Prepare three replicate samples at about 50% of the impurity/degradant limit for each impurity.	

Prepare three replicate samples at about 75% of the impurity/degradant limit for each impurity.

Prepare three replicate samples at about 100% of the impurity/degradant limit for each impurity.

Prepare three replicate samples at about 150% of the impurity/degradant limit for each impurity.

Inject each sample three times and analyze according to the analytical method, adequately bracketed by standard.

Inject samples from the lowest concentration to the highest concentration.

Calculate the % RSD for each individual weight at each level.

Plot the analyte concentration for each individual weight versus the signal response (average of each set of injections).

Perform linear regression analysis, but do not include the origin as a point made and do not force the line through the origin.

Plot the sign and magnitude of the residuals versus analyte concentration.

Check residual plot for outlying values and curvature.

Evaluate y intercept to determine if there is significant departure from zero.

Calculate the recovery for each individual sample weight (average of the three injections).

(Continued)

Category II (Impurity and Degradants) Methods *(Continued)*

Analytical Performance Characteristic or Validation Test	Recommended Procedure	Recommended Acceptance Criteria
	Calculate the average recovery of the three sample weights at each concentration level.	
	Notes: LOD and LOQ experiments should be conducted prior to the accuracy experiments.	
	The impurities/degradants range should span from the LOQ to 150% of the impurities/degradants specification.	
	Separate recovery studies should be performed for each formulation type of various dosage forms (i.e., tablets/capsules, oral/sublingual, etc.).	
	If impurity is already present in the unspiked sample matrix, make three blank samples and inject them with the other 15.	
	Correct spiked sample areas by subtracting the blank, if necessary.	
2. Precision (2 parts):		
Repeatability (method precision)	Prepare six replicate samples solutions from the same assay composite sample according to the analytical method.	The % RSD of the related compounds recovery values should not be greater than 15%.
	If the impurities are not present in significant amounts (>0.2%), spike each sample with impurities at a suitable level.	
	Analyze the samples according to the analytical method, make two injections of each sample.	

118

Calculate the assay results (% recovery) for each sample.

Calculate the individual impurities results (% recovery) for each sample.

Calculate the mean and relative standard deviations (% RSD) of the six sample preparations for assay and impurities.

<table>
<tr><td>Intermediate precision</td><td>

With the following restrictions, have a second analyst execute the following steps:

Perform the work on different days.

Perform the work using different operating conditions and different instruments when possible (e.g., column, apparatus, reagents).

If possible use a different manufacturer's instrument.

Prepare six replicate sample solutions from the same assay composite sample according to the analytical method.

If the impurities are not present in significant amounts (>0.2%), spike each sample with impurities at a suitable level.

Analyze the samples according to the analytical method and make two injections of each sample.

Calculate the assay results (% recovery) for each sample.

Calculate the individual impurities results (% recovery) for each sample.

</td><td>

The % RSD of the impurities/degradants generated on the second day should not be greater than 15%.

The % RSD of the combined assay/recovery values generated by both over both days should not be greater than 15%.

</td></tr>
</table>

(Continued)

119

Category II (Impurity and Degradants) Methods *(Continued)*

Analytical Performance Characteristic or Validation Test	Recommended Procedure	Recommended Acceptance Criteria
	Calculate the average total percent impurities/degradants in each sample or recovery values and the % RSD of all the individual sample weights over both days.	
	Notes: Precision should be investigated with homogenous samples.	
	If it was not possible to obtain homogenous samples, use artificially prepared samples or samples solutions.	
	If the method is used for a range of dosage types (tablets/capsules, oral/sublingual, etc.), the method should be validated for method precision with each dosage type.	
	The same grind method should be used by both analysts if a grind was necessary for sample preparation (e.g., tablets).	
	If there are multiple potencies, precision method should be performed on the low and high potency.	
	If there is only one strength of a product, method precision method should be performed on two different lots of that one strength.	
	A chromatogram of an impurity typical standard, a typical system suitability standard, a typical sample, and a sample spiked with impurities should be included in the validation report.	

3. **Specificity**

In cases where the impurity quantities are limited, the same stock impurity solution can be used by both analysts.

Make an injection of a blank/diluent during each chromatographic run to ensure there are no interferences.

If an internal standard was used, inject the internal standard by itself to confirm specificity.

Prepare a placebo for each potency. If the formulations are dose proportional, placebo interference may be performed on only one potency.

Inject each placebo preparation twice.

Confirm that no peaks can be attributed to the blank/diluent or placebo.

Define any peaks observed by RRT indexed to the active component.

Prepare and inject twice, samples of each individual impurity at the impurity/degradant (e.g., related compound) specification limit.

Prepare two separate spiked solutions containing the active at 100% and each impurity (from two separate impurity preparations) at the limit.

Inject the spiked samples twice to confirm specificity.

Calculate the relative response factor (RRF) for each impurity using the spiked solution preparation, where RRF is the slope of active/slope of impurity.

Confirm that the impurities/degradants do not elute in the elution zone of the active, placebo, or solvent front.

Confirm that the all impurities and degradants are well separated from each other.

(Continued)

121

Category II (Impurity and Degradants) Methods *(Continued)*

Analytical Performance Characteristic or Validation Test	Recommended Procedure	Recommended Acceptance Criteria
	Notes: Primary reference standards should be used if at all possible.	
	For all known impurities/degradants, classify each impurity as process impurity or degradant. This may be done using information from the literature or information provided by the drug substance manufacturer.	
4. Detection limit	4. Detection limit Prepare standard solution and stock solution prepared for each impurity/degradant and the active.	The limit of detection is the first concentration at which the analyte has a S/N ratio of 3:1.
	Prepare a combined sample created by mixing the impurities and active into one solution.	The limit of quantitation is the level at which a S/N ratio of 10:1 is obtained, and the difference between injections is less than 10%.
	Perform serial dilutions to create solutions which span from 150% of the impurity/degradant specification to 0.005% (area percent relative to main active peak). Suggested levels should include:	
	0.005%	
	0.01%	
	0.02%	
	0.05%	
	0.1%	
	0.2%	
	0.5%	

1.0%

1.5%

Note: This test may be done for each compound individually if necessary.

Make five injections of each concentration, adequately bracketed by the standard.

Inject sample solutions from the lowest concentration to the highest concentration.

Calculate the % RSDs at each concentration.

Calculate the S/N ratio for each peak in each sample at each level.

The limit of detection is the first concentration at which the analyte has a S/N ratio of 3:1.

Notes:
The purest analyte available should be used to determine the limit of detection and limit of quantitation (i.e., primary standard such as USP or EPCRS).

5. **Quantitation limit**   The quantitation limit is determined concurrently with the detection limit.

The limit of quantitation is the level at which a S/N ratio of 10:1 is obtained, and the difference between injections is less than 10%.

6. **Linearity**   Prepare standard solutions for the active and a stock standard solution for each impurity

Combine the impurities and active into one solution.

Coefficient of determination (r^2) should be greater than 0.995.

There should be no curvature in the residuals plot.

(Continued)

123

Category II (Impurity and Degradants) Methods (*Continued*)

Analytical Performance Characteristic or Validation Test	Recommended Procedure	Recommended Acceptance Criteria
	Perform serial dilutions to create solutions which span from 150% of the impurity/degradant limit to the limit of quantitation (LOQ) for the active and each impurity. Levels should include:	The *y* intercept should not significantly depart from zero (e.g., area response of *y* intercept should be less than 5% of the response of the midrange concentration value).
	A solution with concentration of the LOQ	
	A solution with concentration of 25% of the impurity/degradant limit	
	A solution with concentration of 50% of the impurity/degradant limit	
	A solution with concentration of 75% of the impurity/degradant limit	
	A solution with concentration of 100% of the impurity/degradant limit	
	A solution with concentration of 150% of the impurity/degradant limit	
	Note: Test may be done for each compound individually. Make five injections of each concentration, adequately bracketed by the standard.	
	Inject solutions from lowest concentration to highest.	
	Calculate % RSD at each concentration.	

Plot the analyte concentration for each set of dilutions separately versus the signal response (average of each set of injections).

Perform linear regression analysis, but do not include the origin as a point and do not force the line through the origin.

Plot the sign and magnitude of the residuals versus analyte concentration.

Check residual plot for outlying values and curvature.

Evaluate y intercept to determine if there is significant departure from zero.

Notes:
The results of the detection limit and quantitation limit may be used for this experiment if appropriate

Coefficient of determination (r^2) should be greater than 0.995.

There should be no curvature in the residuals plot.

The y intercept should not significantly depart from zero (e.g., area response of y intercept should be less than 5% of the response of the nominal 100% concentration value).

7. Range

If not using the samples and results from the accuracy (recovery) experiments perform the following:

Prepare 15 samples by weighing an appropriate amount of placebo.

Spike the placebo samples with 100% of the active sample concentration.

Spike three of each of the these samples with all impurities covering the following ranges:

A solution with concentration of the LOQ

A solution with concentration of 25% of the impurity/degradant limit

A solution with concentration of 50% of the impurity/degradant limit

A solution with concentration of 75% of the impurity/degradant limit

(Continued)

125

Category II (Impurity and Degradants) Methods   (*Continued*)

Analytical Performance Characteristic or Validation Test	Recommended Procedure	Recommended Acceptance Criteria
	A solution with concentration of 100% of the impurity/degradant limit	
	A solution with concentration of 150% of the impurity/degradant limit	
	Note: Test may be done for each compound individually	
	Inject each sample three times and analyze according to the analytical method, adequately bracketed by standard.	
	Inject samples from the lowest concentration to the highest concentration.	
	Plot the analyte concentration for each individual weight versus the signal response (average of each set of injections).	
	Calculate the % RSD for each individual weight at each level.	
	Perform linear regression analysis, but do not include the origin as a point and do not force the line through the origin.	
	Plot the sign and magnitude of the residuals versus analyte concentration.	
	Check residual plot for outlying values and curvature.	
	Evaluate *y* intercept to determine if there is significant departure from zero.	
	Notes: If the impurity is already present in the unspiked sample matrix, make three blank samples and analyze them according to the analytical method.	

If there are multiple potencies, perform range experiments for both the lowest and highest potency.

If the potencies are dose proportional, then the linearity may have been performed using one potency.

If compounds are of low solubility, prepare dry spiked placebos or dissolve the compound in a small amount of solvent in which it is soluble.

The range samples may also be used for accuracy/recovery.

8. **Robustness testing**

Prepare a system suitability standard.

Prepare standard solution containing active at the 100% level spiked with known impurities/degradants at the limit.

Mobile phase variation

Using the mobile phase specified in the test method:

Increase each [major] component by 5%, inject the system suitability standard twice, inject the spiked active sample twice, and measure the appropriate figures of merit.

Decrease each [major] component by 5%, inject the system suitability standard twice, inject the spiked active sample twice, and measure the appropriate figures of merit.

Increase each [major] component by 10%, inject the system suitability standard twice, inject the spiked active sample twice, and measure the appropriate figures of merit.

Measure appropriate figures of merit (e.g., resolution, tailing factor, theoretical plates, and capacity factor) for each variation experiment.

Determine the suitability of the method under each modification determined by taking into account resolution, peak shape, retention time, system pressure, and system suitability parameters.

Determine which system suitability parameters are important to the overall function of the method.

Establish limits for critical parameters.

For column variability ensure all columns used in validation are commercially available.

Ensure that three columns from at least two different lots of packing material are obtained and used.

(Continued)

Category II (Impurity and Degradants) Methods *(Continued)*

Analytical Performance Characteristic or Validation Test	Recommended Procedure	Recommended Acceptance Criteria
	Decrease each [major] component by 10%, inject the system suitability standard twice, inject the spiked active sample twice, and measure the appropriate figures of merit.	Ensure that a brand new column and an old column (>500 injections) are obtained and used.
	Increase each [minor] component by 15%, inject the system suitability standard twice, inject the spiked active sample twice, and measure the appropriate figures of merit. (*Note:* Minor components are those with less than 10 mL/L.)	Ensure the retention times are similar on each of the three columns.
	Decrease each [minor] component by 15%, inject the system suitability standard twice, inject the spiked active sample twice, and measure the appropriate figures of merit.	
	Increase each [minor] component by 30%, inject the system suitability standard twice, inject the spiked active sample twice, and measure the appropriate figures of merit.	
	Decrease each [minor] component by 30%, inject the system suitability standard twice, inject the spiked active sample twice, and measure the appropriate figures of merit.	
	HPLC column temperature variation	
	Using a column heater:	
	Inject the system suitability standard injected twice and inject the spiked active sample twice at the temperature stated in the method, and measure the appropriate figures of merit.	

Inject the system suitability standard injected twice and inject the spiked active sample twice at $+5°$ above stated method temperature and measure the appropriate figures of merit.

Inject the system suitability standard injected twice and inject the spiked active sample twice at $-5°$ above stated method temperature, and measure the appropriate figures of merit.

Mobile phase flow-rate variation

Inject the system suitability standard twice and inject the spiked active sample twice at a 10% increase in flow rate, and measure the appropriate figures of merit.

Inject the system suitability standard injected twice and inject the spiked active sample twice at a 10% decrease in flow rate, and measure the appropriate figures of merit.

Inject the system suitability standard injected twice and inject the spiked active sample twice at a 25% increase in flow rate, and measure the appropriate figures of merit.

Inject the system suitability standard injected twice and inject the spiked active sample twice at a 25% decrease in flow rate, and measure the appropriate figures of merit.

For buffer pH variation

Increase the mobile phase 0.25 pH units, inject system suitability standard injected twice, and inject the spiked active sample twice and measure the appropriate figures of merit.

(Continued)

Category II (Impurity and Degradants) Methods (Continued)

Analytical Performance Characteristic or Validation Test	Recommended Procedure	Recommended Acceptance Criteria
	Decrease the mobile phase 0.25 pH units, inject system suitability standard injected twice, and inject the spiked active sample twice and measure the appropriate figures of merit.	
	HPLC column variation	
	Using three columns from at least two different lots of packing material obtained, inject the system suitability standard injected twice and inject the spiked active sample twice on each column, and measure the appropriate figures of merit.	
	Using a brand new column and an old column (>500 injections), inject system suitability standard injected twice on each column and inject the spiked active sample twice, and measure the appropriate figures of merit.	
	Notes:	
	Relative retention times should be calculated and used to evaluate the effect of method changes on known impurities/degradants.	
9. System suitability determination	Where applicable, calculate system suitability (capacity factor, tailing factor, resolution, theoretical plates, and reproducibility) for each chromatographic run during methods validation.	The following minimum criteria are suggested: $k' \geq 2.0$, where k' is capacity factor
	Identify upper and lower limits for each parameter by analyzing values obtained throughout the validation.	$T \leq 2$, where T is tailing factor
	Establish minimum criteria from this range.	$R > 1.5$, where R is resolution

130

$N \geq 1000$ plates, where N is theoretical plates per column

$\% \text{RSD} \leq 2.0\%$, where $\%$ RSD is the percent relative standard deviation

10. **Forced degradation studies**

The following degradation conditions are recommended as a starting point:

Degradation Reaction	Typical Conditions
Acid hydrolysis	Sample in aqueous acid or acidified solvent at ~0.5 N up to 24 hours (or)
	Heat/reflux or UV radiation in ~0.5 N HCl up to 24 hours.
Base hydrolysis	Sample in aqueous base or basic solvent at ~0.5 N up to 24 hours (or)
	Heat/reflux or UV radiation in ~0.5 N NaOH up to 24 hours.
Oxidation	Treat with ~3% H_2O_2 up to 24 hours (or)
	UV irradiation in ~3% H_2O_2 up to 24 hours.
Light decomposition (photolysis)	Expose to high-intensity UV light in suitable increments, up to 24 hours.
Thermal decomposition (pyrolysis)	Expose to ~100° C heat in suitable increments, up to 24 hours.

Confirm peak purity of the degraded spiked placebo sample main peak using UV or MS analysis.

Ensure peak purity is performed for each degradation condition and that the samples have degraded sufficiently or degradation had stopped.

Evaluate spectral overlay of the sufficiently degraded impurities/degradants spiked placebos in addition to peak purity to demonstrate that the degradants are resolved from the analyte.

Determine relative retention times of the degradants.

Assess chromatograms of the sufficiently degraded spiked placebo overlaid with the degraded placebo under each degradation.

(*Continued*)

131

Category II (Impurity and Degradants) Methods (*Continued*)

Analytical Performance Characteristic or Validation Test	Recommended Procedure	Recommended Acceptance Criteria
	Obtain or prepare solutions of ~0.5 N HCL, ~0.5 N NaOH, ~3% H_2O_2, and purified water.	
	Prepare blank samples for each condition, including light and heat.	
	Prepare a standard, spiked placebo, and unspiked placebo as appropriate in preparation for degradation.	
	Analyze samples and blanks according to the method outlined in the validation prior to degradation, establishing an initial time point.	
	Expose these standards, spiked placebos, and an unspiked placebo to acid, base, oxidation, heat, and light using the condition guidelines listed above.	
	Neutralize acid and base samples prior to analysis by pipetting in an amount of acid or base solution equal to the sample aliquot, and then diluting to volume with water, diluent, or mobile phase.	
	Analyze samples and blanks at varying intervals over 24 hours, according to the method outlined in the validation following degradation. Assess whether sufficient degradation (10%–30%) has occurred.	
	Calculate the percent recovery of the appropriate solutions to determine the extent of degradation.	

(Continued)

If the acid, base, and peroxide solutions were not sufficiently degraded in 24 hours, the acid, base, and peroxide solutions should have been exposed to heat/light until at least 10%–30% degradation is achieved or 24 hours of exposure has elapsed. Moreover, if the samples are overdegraded, lessening the length and severity of the conditions to obtain the 10%–30% is acceptable.

Perform peak purity analysis on the main analyte peak using diode array and/or mass spectra analysis.

Capture and display chromatograms for each degradation on appropriately degraded (10%–30%) samples.

Capture and display peak purity analyses for each degradation condition.

Determine whether the method is specific for the degraded samples.

Notes:

Forced degradation studies are designed to produce potential degradation products which may be encountered in real-world scenarios. Degradants generated may or may not be what is seen during stability studies.

Performing the actual degradation in these studies is not an exact science and may require modifying conditions to obtain the desired 10%–30% degradation. However, if the maximum conditions listed here do not produce degradation, then it is not necessary to continue the experiments until degradation occurs. A statement is then added to the report confirming the stable nature of the molecules.

Category II (Impurity and Degradants) Methods *(Continued)*

Analytical Performance Characteristic or Validation Test	Recommended Procedure	Recommended Acceptance Criteria
	For multicomponent formulations (e.g., with more than one active), individual active solutions should be made for each component. For product families that utilize the same excipients, forced degradation should be formed on only one formulation.	
	If dyes are utilized in the product information, placebos for each different formulation can be prepared and used for the forced degradation study. At the discretion of the front line supervisor, if dyes are utilized in the product formulation, a worst case placebo can be used for the forced degradation study.	
	Acid and base samples should be prepared at 2× the nominal concentration, so that they may be neutralized prior to analysis.	
	If the analyte is not soluble in aqueous solutions, a small amount of suitable organic solvent may be used to dissolve the sample.	
	The acid, base, and oxidizing solutions may be made up in solvent.	
	The concentration of the acid, base, and peroxide solutions may be reduced or the sample concentration may be lowered with subsequent increase of the injection volume to maintain the appropriate amount of material on column.	

If deemed necessary by the supervisor, forced degradation studies may also be performed on dry powder samples.

If forced degradation demonstrates lack of specificity, analysis of expired finished product may be used to prove that the forced degradation conditions are not producing real degradant peaks.

Targeted recovery for the initial samples prior to degradation should be in the 95%–105% range.

The run time for forced degradation samples should be sufficiently long to observe the retention time of the latest eluting active or degradant.

For product families that utilize the same excipient, forced degradation should be formed on only one formulation.

11. **Solution stability studies**

Prepare fresh standard and spiked placebo (or actual tablet/capsule) as per the test method. Ensure that both stock and working solutions are available for analysis.

Analyze these solutions analyzed as per the test method, establishing a time zero value for each.

Place an aliquot of each solution in clear glassware and expose to ambient (benchtop) conditions.

Place an aliquot of each solution placed in amber glassware and expose to ambient (benchtop) conditions.

For assay-level standards, the fresh standard and the verification standard should not differ more that 2.0%.

For low- (impurity, 1%–2% of method concentration) level standards, the fresh standard and fresh verification standard should not differ by more than 2.0%.

For the assay level, the standard and sample solutions are considered sufficiently stable over time if the recovery value does not vary more than ±1.5% (absolute) from the initial result.

(Continued)

Category II (Impurity and Degradants) Methods (Continued)

Analytical Performance Characteristic or Validation Test	Recommended Procedure	Recommended Acceptance Criteria
	Place an aliquot of each solution in clear glassware and place in a refrigerator.	For low- (impurity) level standards, the solution is considered sufficiently stable over time if the recovery value does not vary more than ±3.0% (absolute) from the initial result.
	Analyze these samples versus fresh standard every 24 hours for at least two (2) days.	
	Make two injections from each solution.	For impurities/degradants methods, the samples are considered to be sufficiently stable over time if the impurities/degradants level does not vary more than 0.2% (absolute) from the initial sample analysis.
	Calculate the percent recoveries calculated for all solutions.	
	Notes:	
	After two days, samples may be evaluated at intervals at the analyst's discretion.	
	A fresh standard and fresh verification standard should be prepared each day.	
12. Filter retention studies	Prepare three placebos for the lowest potency, spiked with dry active at 100% of the nominal assay concentration, and each of the impurities at the limits.	For an acceptable filter, the difference between the filtered sample and the centrifuged sample for impurities is NMT 5% (absolute).
	Where necessary, three product sample preparations (tablets, capsules, drug substance, etc.) containing the known impurities in sufficient quantities (~0.5%) may be substituted for the dry spiked placebos.	The individual and average recovery of each impurity should be within the range of 75%–125%.

136

Filter portions of each individual spiked placebo sample filtered through at least two (2) prospective filters. Ensure the filters are commercially available. If this is a revalidation, one of the prospective filters must be one that is specified in the respective monograph or test method.

Centrifuge an aliquot of each individual spiked placebo (e.g., 3000 rpm or 10 minutes suggested).

Inject all three samples three times each and analyze according to the analytical method adequately bracketed by the standard.

Calculate the recovery of each individual impurity for each sample weight and % RSD of the replicate injections.

Calculate the average recovery for each filter.

Calculate the percent difference of the average result for each filter versus the average result of the centrifuged sample for each impurity.

Notes:
If the impurities are already present in the unspiked sample matrix, prepare three blank samples as well.

Subtract the impurities areas of the blank if necessary in the calculations.

13. Extraction efficiency studies

Not normally performed for Category II methods (impurities and degradants).

Note: These studies are not normally performed due to the difficulty in obtaining aged samples with sufficient impurities and degradants present.

ATTACHMENT V: RECOMMENDED EXPERIMENTAL PROCEDURES AND ACCEPTANCE CRITERIA FOR CATEGORY III METHODS (DISSOLUTION)

Category III Methods (Dissolution)

Analytical Performance Characteristic or Validation Test	Recommended Procedure	Recommended Acceptance Criteria
1. Accuracy (recovery)	**If not using the samples and results from the range experiments perform the following:**	The percent recovery of the spiked placebos should be within $100 \pm 2\%$ for the average of each set of three weights.
	Prepare 15 samples by weighing an appropriate amount of placebo (the placebo remains 100% of method concentration in all samples), with respect to the concentration specified in the method being validated.	Each individual sample recovery should lie within the range of 97%–103%.
	Spike the placebo samples with one stock active solution (using diluent as per the sample preparation) at each of the five levels. The five levels should span the range of (Q–25%) to 120% of the nominal drug product strength.	Coefficient of determination (r^2) should be greater than 0.997.
	Prepare three (3) replicate weights for each level.	There should be no curvature in the residuals plot.
	Inject each sample three times and analyze according to the analytical method, adequately bracketed by standard.	The y intercept should not significantly depart from zero (e.g., area response of y intercept should be less than 5% of the response of the nominal 100% concentration value).
	Inject samples from the lowest concentration to the highest concentration.	
	Calculate the % RSD for each individual weight at each level.	
	Plot the analyte concentration for each individual weight versus the signal response (average of each set of injections).	

Perform linear regression analysis, but do not include the origin as a point and do not force the line through the origin.

Plot the sign and magnitude of the residuals versus analyte concentration.

Check residual plot for outlying values and curvature.

Evaluate y intercept to determine if there is significant departure from zero.

Calculate the recovery for each individual sample weight (average of the three injections).

Calculate the average recovery of the three sample weights at each concentration level.

Notes:
Separate recovery studies should be performed for each formulation type of various dosage forms (i.e., tablets/capsules, oral/sublingual, etc.)

2. Precision (2 parts):

Repeatability (method precision)	Perform dissolution on 6 individual dosage units from both the lowest potency and the highest potency, according to the dissolution method. If there is only one potency, perform dissolution using two lots of that potency.	The % RSD should be NMT 3.0%.
	Analyze the samples according to the analytical method and make two injections of each sample.	
	Calculate the percentage dissolved of each individual unit.	

(*Continued*)

Category III Methods (Dissolution)   (*Continued*)

Analytical Performance Characteristic or Validation Test	Recommended Procedure	Recommended Acceptance Criteria
	Calculate the mean and relative standard deviations (% RSD) of the six dissolution results.	The % RSD should be NMT 3.0%
Intermediate precision	**With the following restrictions, have a second analyst execute the followings steps:**	The % RSD for the combined two results should be NMT 4.0%.
	Perform the work on different days.	
	Perform the work using different operating conditions and different instruments when possible (e.g., column, apparatus, reagents).	
	If possible use a different manufacturer's dissolution apparatus and HPLC instrument.	
	Perform dissolution on 6 individual dosage units from both the lowest potency and the highest potency, according to the dissolution method. If there is only one potency, perform dissolution using two lots of that potency.	
	Analyze the samples according to the analytical method and make two injections of each sample.	
	Calculate the percentage dissolved of each individual unit.	
	Calculate the mean and relative standard deviations (% RSD) of the six dissolution results.	

Notes:

Precision should be investigated with homogenous samples.

If the method is used for a range of dosage types (tablets/capsules, oral/sublingual, etc.), the method should be validated for method precision with each dosage type.

If there are multiple potencies, precision method should be performed on the low and high potency.

If there is only one strength of a product, method precision should be performed on two different lots of that one strength.

3. Specificity

Prepare and inject twice a representative sample of dissolution media.

Prepare and inject twice a blank and diluent (if appropriate) sample.

If an internal standard was used, prepare and inject twice the internal standard by itself to confirm specificity.

Prepare a placebo for each potency. If the formulations are dose proportional, placebo interference may be performed on only one potency.

For capsule products, dissolve a capsule shell in an appropriate volume of dissolution media.

Inject the capsule shell sample twice.

Inject each placebo preparation twice.

Confirm that no interference in the elution zone of the active occurred from the dissolution media, blank/diluent, internal standard (if applicable), capsule shell (if applicable), or the placebo.

(Continued)

141

Category III Methods (Dissolution)   *(Continued)*

Analytical Performance Characteristic or Validation Test	Recommended Procedure	Recommended Acceptance Criteria
	Confirm that no peaks can be attributed to the dissolution media, blank/diluent placebo, or capsule shell samples.	
	Define any peaks observed by RRT indexed to the active component.	
	Notes: If there is significant interference from the capsule shell, it can be corrected by preparing a capsule shell blank as per USP <711>.	
4. Detection limit	Not performed for Category III methods (dissolution).	
5. Quantitation limit	Not performed for Category III methods (dissolution).	
6. Linearity	Prepare five standard solutions of the analyte at ~50%, 70%, 100%, 130%, and 150% of the nominal method concentration using serial dilutions from a stock solution.	Coefficient of determination (r^2) should be greater than 0.9999.
	Make three injections at each concentration, adequately bracketed by the standard.	There should be no curvature in the residuals plot.
	Make sure to inject samples from the lowest concentration to the highest concentration to reduce the effects, if any, of carryover from the higher concentration samples.	The y intercept should not significantly depart from zero. (e.g., area response of y intercept should be less than 5% of the response of the midrange concentration value).
	Calculate the % RSD at each concentration.	

Plot the analyte concentration for each set of dilutions separately versus the signal response (average of each set of injections).

Perform linear regression analysis, but do not include the origin as a point made and do not force the line through the origin.

Plot the sign and magnitude of the residuals versus analyte concentration.

Check residual plot for outlying values and curvature.

Evaluate y intercept to determine if there is significant departure from zero.

Notes:
If method is to be used for multiple analyte concentrations, ensure linearity was examined from 50% of the lowest nominal concentration to 150% of highest nominal concentration.

Coefficient of determination (r^2) should be greater than 0.997.

There should be no curvature in the residuals plot.

The y intercept should not significantly depart from zero (e.g., area response of y intercept should be less than 5% of the response of the nominal 100% concentration value).

7. **Range**

If not using the samples and results from the accuracy (recovery) experiments perform the following:

Prepare 15 samples by weighing an appropriate amount of placebo (the placebo remains 100% of method concentration in all samples) with respect to the concentration specified in the method being validated.

Spike the placebo samples with one stock active solution (using diluent as per the sample preparation) at each of the five levels. The five levels should span the range of (Q–25%) to 120% of the nominal drug product strength.

Prepare three (3) replicate weights for each level.

(Continued)

Category III Methods (Dissolution) (Continued)

Analytical Performance Characteristic or Validation Test	Recommended Procedure	Recommended Acceptance Criteria
	Inject each sample three times and analyze according to the analytical method, adequately bracketed by standard.	
	Inject samples from the lowest concentration to the highest concentration.	
	Calculate the % RSD for each individual weight at each level.	
	Plot the analyte concentration for each individual weight versus the signal response (average of each set of injections).	
	Perform linear regression analysis, but do not include the origin as a point made and do not force the line through the origin.	
	Plot the sign and magnitude of the residuals versus analyte concentration.	
	Check residual plot for outlying values and curvature.	
	Evaluate y intercept to determine if there is significant departure from zero.	
	Calculate the recovery for each individual sample weight (average of the three injections).	
	Calculate the average recovery of the three sample weights at each concentration level.	

Notes:

If there are multiple potencies, perform range experiments for both the lowest and highest potency.

If the potencies are dose proportional, then the linearity may have been performed using one potency.

If compounds are of low solubility, prepare dry spiked placebos or dissolve the compound in a small amount of solvent in which it is soluble.

The range samples may also be used for accuracy/recovery.

The range of (Q–25%) to 120% is best described in an example. For instance, if the specification for a controlled-release product covers a range from 25%, after 1 hour, and up to 90% after 24 hours, then the range to be validated is from 0% to 110% of label claim.

8. Robustness testing

Mobile phase variation

Using the mobile phase specified in the test method:

Increase each [major] component by 5%, inject the system suitability standard twice, and measure the appropriate figures of merit.

Decrease each [major] component by 5%, inject the system suitability standard twice, and measure the appropriate figures of merit.

Increase each [major] component by 10%, inject the system suitability standard twice, and measure the appropriate figures of merit.

Measure appropriate figures of merit (e.g., resolution, tailing factor, theoretical plates, and capacity factor) for each variation experiment.

Determine the suitability of the method under each modification determined by taking into account peak shape, retention time, system pressure, and system suitability parameters.

Determine which system suitability parameters are important to the overall function of the method.

(Continued)

Category III Methods (Dissolution) (Continued)

Analytical Performance Characteristic or Validation Test	Recommended Procedure	Recommended Acceptance Criteria
	Decrease each [major] component by 10%, inject the system suitability standard twice, and measure the appropriate figures of merit.	Established limits for critical parameters.
	Increase each [minor] component by 15%, inject the system suitability standard twice, and measure the appropriate figures of merit. (Note: Minor components are those with less than 10 mL/L.)	For column variability, ensure all columns used in validation are commercially available.
	Decrease each [minor] component by 15%, inject the system suitability standard twice, and measure the appropriate figures of merit.	Ensure that three columns from at least two different lots of packing material are obtained and used.
	Increase each [minor] component by 30%, inject the system suitability standard twice, and measure the appropriate figures of merit.	Ensure that a brand new column and an old column (>500 injections) are obtained and used.
	Decrease each [minor] component by 30%, inject the system suitability standard twice, and measure the appropriate figures of merit.	Ensure the retention times are similar on each of the three columns.
HPLC column temperature variation		
	Using a column heater:	
	Inject the system suitability standard injected twice at the temperature stated in the method, and measure the appropriate figures of merit.	
	Inject the system suitability standard injected twice at +5° above stated method temperature and measure the appropriate figures of merit.	

(*Continued*)

Inject the system suitability standard injected twice at −5° above stated method temperature and measure the appropriate figures of merit.

Mobile phase flow-rate variation

Inject the system suitability standard twice at a 10% increase in flow rate and measure the appropriate figures of merit.

Inject the system suitability standard injected twice at a 10% decrease in flow rate, and measure the appropriate figures of merit.

Inject the system suitability standard injected twice at a 25% increase in flow rate, and measure the appropriate figures of merit.

Inject the system suitability standard injected twice at a 25% decrease in flow rate, and measure the appropriate figures of merit.

For buffer pH variation

Increase the mobile phase 0.25 pH units, inject system suitability standard injected twice, and measure the appropriate figures of merit.

Decrease the mobile phase 0.25 pH units, inject system suitability standard injected twice, and measure the appropriate figures of merit.

HPLC column variation

Using three columns from at least two different lots of packing material obtained, inject the system suitability standard injected twice on each column, and measure the appropriate figures of merit.

Category III Methods (Dissolution) *(Continued)*

Analytical Performance Characteristic or Validation Test	Recommended Procedure	Recommended Acceptance Criteria
	Using a brand new column and an old column (>500 injections), inject system suitability standard injected twice on each column, and measure the appropriate figures of merit.	
9. System suitability determination	Where applicable, calculate system suitability (capacity factor, tailing factor, resolution, theoretical plates, and reproducibility) for each chromatographic run during methods validation.	The following minimum criteria are suggested: $k' \geq 2.0$, where k' is capacity factor $T \leq 2$, where T is tailing factor $R > 1.5$, where R is resolution
	Identify upper and lower limits for each parameter by analyzing values obtained throughout the validation.	$N \geq 1000$ plates, where N is theoretical plates per column
	Establish minimum criteria from this range.	$\% \text{RSD} \leq 2.0\%$, where $\% \text{RSD}$ is the percent relative standard deviation
10. Forced degradation studies	Not performed for Category IIII methods (dissolution).	
11. Solution stability studies	Prepare fresh standard as per the test method. Ensure that both stock and working solutions are available for analysis.	For assay-level standards, the fresh standard and the verification standard should not differ more that 1.5%.

Prepare or obtain dissolution samples. (e.g., drug product dissolved in dissolution media and diluted as appropriate).

Analyze these solutions per the test method, establishing a time zero value for each.

Place an aliquot of each solution in clear glassware and expose to ambient (benchtop) conditions.

Place an aliquot of each solution placed in amber glassware and exposed to ambient (benchtop) conditions.

Place an aliquot of each solution in clear glassware and place in a refrigerator.

Analyze these samples versus fresh standard every 24 hours for at least two (2) days.

Make two injections made of each solution.

Calculate the percent recoveries calculated for all solutions.

Notes:

After two days, samples may be evaluated at intervals at the analyst's discretion.

A fresh standard and fresh verification standard should be prepared each day.

The solutions are considered sufficiently stable over time taken for 2.0% decomposition as measured by loss of analyte.

(Continued)

149

Category III Methods (Dissolution)   (Continued)

Analytical Performance Characteristic or Validation Test	Recommended Procedure	Recommended Acceptance Criteria
12. Filter retention studies	Prepare three placebos for the lowest potency, spiked with dry active at 100% of the nominal assay concentration.	For the spiked placebos, the percentage recovery should be $100 \pm 2.0\%$ for the average of each set of three weights.
	Where necessary, three product sample preparations (tablets, capsules, drug substance, etc.) may be substituted for the dry spiked placebos.	For each individual spiked placebo the recovery should be 97%–103%.
	Dissolve in dissolution media to obtain the nominal sample concentration obtained during dissolution analysis.	For an acceptable filter, the difference between the filtered sample and the centrifuged sample NMT 2.0% (absolute).
	Filter portions of each individual spiked placebo sample filtered through at least two (2) prospective filters. Ensure the filters are commercially available. If this is a revalidation, make sure one of the prospective filters specified in the respective monograph or test method is used.	
	Centrifuge an aliquot of each individual spiked placebo (e.g., 3000 rpm or 10 minutes suggested)	
	Inject all three samples three times each, and analyze according to the analytical method adequately bracketed by the standard.	
	Calculate the recovery of each individual sample weight and % RSD of the replicate injections.	
	Calculate the average recovery for each filter.	
	Calculate the percent difference of the average result for each filter versus the average result of the centrifuged sample.	
13. Extraction efficiency studies	Not performed for Category III methods (dissolution).	

150

ATTACHMENT VI: RECOMMENDED EXPERIMENTAL PROCEDURES AND ACCEPTANCE CRITERIA FOR CATEGORY IV METHODS (IDENTIFICATION)

Category IV Methods (Identification)

Analytical Performance Characteristic or Validation Test	Recommended Procedure	Recommended Acceptance Criteria
1. Accuracy (recovery)	Not performed for Category IV methods (identification).	
2. Precision (2 parts):		
Repeatability (method precision)	Not performed for Category IV methods (identification).	
Intermediate precision	Not performed for Category IV methods (identification).	
3. Specificity	Make an injection of a blank/diluent during each chromatographic run to ensure there are no interferences.	Confirm that no interference in the elution zone of the active occurred from the blank/diluent, internal standard (if applicable) or the placebo.
	If an internal standard was used, inject the internal standard by itself to confirm specificity.	
	Prepare a placebo for each potency. If the formulations are dose proportional, placebo interference may be performed on only one potency.	
	Inject each placebo preparation twice.	
	Confirm that no peaks can be attributed to the blank/diluent or placebo.	

(Continued)

151

Category IV Methods (Identification) *(Continued)*

Analytical Performance Characteristic or Validation Test	Recommended Procedure	Recommended Acceptance Criteria
	Define any peaks observed by RRT indexed to the active component.	
	Notes: The discrimination of a procedure may be confirmed by obtaining positive results (perhaps by comparison with a known reference material) from samples containing the analyte, coupled with negative results from samples that do not contain the analyte.	
4. Detection limit	Not performed for Category IV methods (identification).	
5. Quantitation limit	Not performed for Cateegory IV methods (identification).	
6. Linearity	Not performed for Category IV methods (identification).	
7. Range	Not performed for Category IV methods (identification).	
8. Robustness testing	Not performed for Category IV methods (identification).	
9. System suitability determination	Not performed for Category IV methods (identification).	
10. Forced degradation studies	Not performed for Category IV methods (identification)	

11. Solution stability studies   Not performed for Category IV methods (identification).

13. Extraction efficiency studies   Not performed for Category IV methods (identification).

Revision History

Version	Effective	Author	Summary of Change
-00	dd/mm/yyyy		New
-01	dd/mm/yyyy		Added............
-02	dd/mm/yyyy		Modified...........

TEMPLATE FOR AN EXAMPLE END-USER REQUIREMENTS QUESTIONNAIRE

END-USER REQUIREMENTS QUESTIONNAIRE

▫ Where will the method be used (e.g., geographic locations and actual facility)?

▫ Who will use the method?

▫ What equipment restrictions may exist?

▫ What supply restrictions may exist?

▫ What expertise restrictions may exist?

▫ What language barriers may exist?

Validating Chromatographic Methods. By David M. Bliesner
Copyright © 2006 John Wiley & Sons, Inc.

□ What USP/ICH methods category is it?

□ What validation characteristics are needed?

□ Have the end users and end users' supervisors been included in the method development process?

□ Have the end users had a chance to review and provide input on the commissioning document?

□ Have the end users had experience with this product or product class in the past?

□ Have the end users been interviewed about their experiences with this product or product class?

□ Do the end users have any special requirements or needs that have not been previously considered?

□ Other consideration for final use of the method:

APPENDIX IV

TEMPLATE FOR AN EXAMPLE METHOD REVIEW CHECKLIST

METHOD REVIEW CHECKLIST

Before beginning the final validation, perform a sanity check of the separation, considering and/or evaluating the following:

- □ Compound solubility

- □ Compound stability in solution

- □ pK_a versus buffer pH

- □ Toxicity of materials

- □ Appropriateness of lambda max (e.g., a true maxima, on slope)

- □ UV cut-off of mobile phase

- □ Aggressiveness of mobile phase on column

Validating Chromatographic Methods. By David M. Bliesner
Copyright © 2006 John Wiley & Sons, Inc.

- Gradient versus isocratic

- Detector choice

- Standard/sample diluent-mobile phase injection match

- Injection volume

- Availability of reagents and supplies (future availability as well as all plant locations)

- Potential excipient compatibility issues (to ascertain future degradant growth)

- Are the concentrations of standard and sample solutions within the dynamic range of the technique?

- Are the dilution schemes applicable to the QC environment or other end-user lab?

- Is there another chromatographic technique which is more appropriate than HPLC (e.g., GC for volatile compounds)?

- Are the types and sizes of volumetric glassware appropriate for the QC environment or other end-user lab?

- Ion pair mobile phase with gradient-potential problems

- What was run previously on chromatographic system? Carryover, or mobile phase incompatibility issues may exist.

- Method to prevent columns which use ion pair mobile phases from being used for anything else.

- Directions for column care at the end of the run. Ion pair should be stored in ion pair if at all possible.

- Concentration of ion pair reagent very high

▫ Why using ion pair when more selective phases are available?

▫ Is stationary phase a good match for the separation?

▫ Is stationary phase suitability for mobile phase used (e.g., CN columns can react)?

▫ Are the samples and standards compatible with the mobile phase and diluents (e.g., can solvent or pH reactions occur with the analyte)?

▫ Can tautomerism exist with the analyte?

▫ Are diluents and blanks similar to mobile phase to prevent injection solvent mismatch?

▫ Concentration of injected solution produces responses which are within the dynamic range of the detector (e.g., both variable wavelength and DAD can produce spectral artifacts if absorbance is too high)?

▫ Is the mobile phase or components of the mobile phase volatile?

▫ Is the mobile phase stable (e.g., is microbial growth a problem)?

▫ Does the mobile phase absorb water from the air? Is this a problem (e.g., normal phase methods)?

▫ Are there any special equipment and glassware cleaning requirements associated with the active component?

APPENDIX V

TEMPLATE FOR AN EXAMPLE STANDARD TEST METHOD

Title:	Effective:		Document No:
Method for the Determination of the Assay and Impurities for Compound A Tablets, 5- and 10-mg, and Compound A API: Example Template	**DD/MM/YYYY**		
	Supersedes:		Page:
Document Type:	Department:		Reference Document:

Prepared by:	Approved by:	Filed by:
Place title here	Place title here	Place title here
Place name here	Place name here	Place name here
Date: dd/mm/yyyy	Date: dd/mm/yyyy	Date: dd/mm/yyyy

I. Reagents

Sodium acetate HPLC grade

Glacial acetic acid, HPLC grade

THF, HPLC grade

Acetonitrile, HPLC grade

Water, in-house deionized water

Compound A reference standard

II. Mobile Phase Preparation

Mobile phase A: 70:24:6 of 25 mM sodium acetate buffer (pH 5.0) Acetonitrile: THF

Mobile phase B: 60:40 THF:acetonitrile

Preparation of 25 mM sodium acetate buffer, pH 5.0

Accurately weigh 9.6 g of sodium acetate and quantitatively transfer to a 5-L container. Add 5 L of purified water and mix until dissolved. Adjust the pH to 5.0 ($\pm$ 0.05) using glacial acetic acid. This solution preparation may be scaled up as necessary.

Preparation of mobile phase A (70:24:6:sodium acetate buffer (pH 5.0):acetonitrile:THF (v:v:v))

Prepare mobile phase A to have a ratio of 70:24:6 sodium acetate buffer (pH 5.0):acetonitrile:THF. For example, to prepare 4 L, combine 2800 mL of sodium acetate buffer (pH 5.0), 960 mL of acetonitrile, and 240 mL of THF. Mix well and degas by sonication prior to use.

Preparation of mobile phase B (60:40::THF:acetonitrile)

Prepare mobile phase B to have a ratio of 60:40 THF:Acetonitrile. For example, to prepare 4 L, combine 2400 mL of THF and 1600 mL of Acetonitrile. Mix well and degas by sonication prior to use.

Diluent (50:50 mobile phase A:THF)

Prepare a solution of mobile phase A and THF with a ratio of 50:50. For example, to prepare 4 L, combine 2000 mL of mobile phase A and 2000 mL of THF. Mix well.

III. Standard Preparation

Assay standard solution (0.2 mg/mL) (prepare in duplicate)

Accurately weigh approximately 20 mg of compound A reference standard into a 100-mL volumetric flask. Add ~50 mL of diluent and sonicate for 10 –15 minutes with intermittent shaking until a clear solution is obtained. Allow solution to cool to room temperature and dilute to volume with diluent. Label as main and check assay standards.

Impurity standard solution (0.0002 mg/mL) (prepare in duplicate)

Transfer 10.0 mL of the assay standard solution into a 1000-mL volumetric flask. Dilute to volume with diluent and mix well. Transfer 10.0 mL of this solution into a 100-mL volumetric flask. Dilute to volume with diluent and mix well. Label as main and check impurity standards.

Sensitivity solution (0.0001 mg/mL)

Transfer 10.0 mL of the assay standard solution into a 1000-mL volumetric flask. Dilute to volume with diluent and mix well. Transfer 5.0 mL of this solution into a 100-mL volumetric flask. Dilute to volume with diluent and mix well.

IV. Sample Preparation

Assay and impurities sample preparation (prepare in duplicate)

Accurately weigh 40 tablets, grind to fine powder, and calculate the average tablet weight. Accurately weigh duplicate samples of ~155 mg (of the powder equivalent to 5 mg or 10 mg of compound A) into separate 50-mL volumetric flasks. Fill approximately three by four full with diluent and sonicate for 10–15 minutes with intermittent shaking. Dilute to volume with diluent and centrifuge or filter a portion of the solution to obtain a clear solution.

Drug substance (API) sample preparation

Accurately weigh ~20 mg of API into a 100-mL volumetric flask. Add ~50 mL of diluent and sonicate for 10–15 minutes with intermittent shaking until a clear solution is obtained. Allow solution to cool to room temperature, and dilute to volume diluent.

V. Chromatographic Conditions

Column:	Serta, Sleeper C4, 3.5 mm, 4.6×150 mm
Column temp:	30°C
Mobile phase A:	70:24:6::25 mM sodium acetate buffer (pH 5.0):acetonitrile:THF (v:v:v)
Mobile phase B:	60:40::THF:acetonitrile (v:v)
Flow rate:	1.50 mL/min
Detector:	UV at 254 nm
Injection volume:	50 µl

Run time: 35 min
Gradient profile: (ramps are linear)

Time (min)	% A	% B
0	100	0
4	100	0
26	51	49
27	100	0
35	100	0

Assay

Assay standards and sample preparations may be analyzed according to the following injection sequence:

Test Solution	# of Injections
Blank (diluent)	2
Main standard	5
Check standard	2
Assay sample 1	1
Assay sample 2	1
Assay sample X	1
Main standard	1

Note: Inject samples with no more than 10 sample injections between bracketing standard solutions.

Impurities

Impurity standards and sample preparations may be analyzed according to the following injection sequence:

Test Solution	# of Injections
Blank (diluent)	2
Sensitivity solution	1
Main impurity standard	6
Check impurity standard	2
Impurity sample 1	1
Impurity sample 2	1
Impurity sample X	1
Main impurity standard	1

Note: Inject samples with no more than 10 sample injections between bracketing standard solutions.

Assay and Impurities

Standards and sample preparations may be analyzed according to the following injection sequence:

Test Solution	# of Injections
Blank (diluent)	2
Sensitivity solution	1
Check impurity standard	2
Main impurity standard	6
Check standard	2
Main standard	5
Main impurity standard	1
Sample 1	1
Sample 2	1
Sample X	1
Main standard	1
Main impurity standard	1

Note: Inject no more than 8 samples between bracketing standards.

VI.   System Suitability

System suitability criteria will be established after completion of the validation. The following minimum criteria should be met for the assay standard.

- $k' \geq 5.0$, where k' is the capacity factor
- $T \leq 2$, where T is the tailing factor
- $R \geq 2.0$, where R is the resolution between adjacent peaks
- $N \geq 15000$, where N is the number of theoretical plates
- % RSD ≤ 2.0, where RSD is the relative standard deviation of five replicate standard injections
- Main: check standard agreement $\pm$ 2.0%

Requirements for the impurity standards are:

- % RSD ≤ 10.0, where % RSD is the percent relative standard deviation of six replicate standard injections
- Main: Check standard agreement $\pm 10.0\%$, required only if run is for impurity testing only. Not needed if assay and impurity testing run together
- S/N ≥ 10 for the compound A peak in the sensitivity solution

VII.   Calculations

Compound A assay for tablets:

$$\text{Compound A (mg/tablet)} = \frac{A_{\text{sam}}}{A_{\text{std}}} \times \frac{\text{Wt}_{\text{std}} \times \text{PF}}{100 \text{ mL}} \times \frac{50 \text{ mL}}{\text{Wt}_{\text{sam}}} \times \text{Wt}_{\text{avg}}$$

$$\text{Compound A (\%label)} = \frac{\text{Compound A (mg/tablet)}}{\text{TS}} \times 100$$

Compound A assay for API:

$$\text{Compound A (\%w/w)} = \frac{A_{\text{sam}}}{A_{\text{std}}} \times \frac{\text{Wt}_{\text{std}} \times \text{PF}}{100 \text{ mL}} \times \frac{100 \text{ mL}}{\text{Wt}_{\text{sam}}} \times 100$$

% Impurity for tablets:

$$\text{\% Impurity (\%w/w)} = \frac{A_{\text{imp}}}{A_{\text{std}}} \times \frac{\text{Wt}_{\text{std}} \times \text{PF}}{\text{Wt}_{\text{sam}}} \times \frac{\text{Wt}_{\text{avg}}}{\text{TS}}$$

$$\times \frac{50 \text{ mL} \times 10 \text{ mL}}{100 \text{ mL} \times 1000 \text{ mL}} \times \text{RRF} \times 100$$

% Impurity for API:

$$\text{\% Impurity (\%w/w)} = \frac{A_{\text{imp}}}{A_{\text{std}}} \times \frac{\text{Wt}_{\text{std}} \times \text{PF}}{\text{Wt}_{\text{sam}}}$$

$$\times \frac{100 \text{ mL} \times 10 \text{ mL}}{100 \text{ mL} \times 1000 \text{ mL}} \times \text{RRF} \times 100$$

where

A_{sam}	=	area response of compound A peak in the sample preparation
A_{imp}	=	area response of an impurity peak in the sample preparation
A_{std}	=	average area response of compound A peak in the standard preparation
A_{istd}	=	average area response of compound A peak in the impurity standard preparation
Wt_{std}	=	weight of standard, mg
Wt_{sam}	=	weight of powder taken in sample, mg
Wt_{avg}	=	average weight of tablets, mg
PF	=	standard purity factor
TS	=	tablet strength, mg
RRF	=	relative response factor (RRF = 1 for unknown impurities)

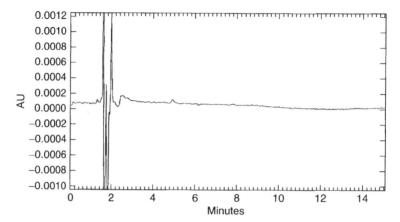

Example of a blank injection.

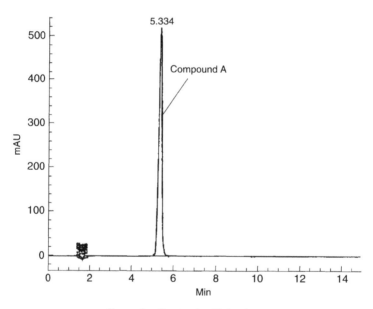

Example of a standard injection.

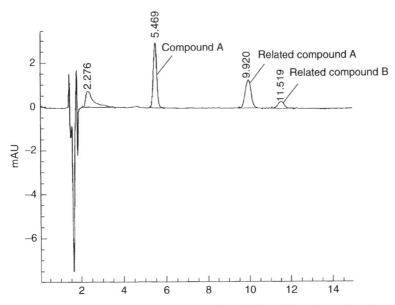

Impurity standard spiked with related A and B to show elution order and retention time of relateds.

Revision History

Version	Effective	Author	Summary of Change
-00	dd/mm/yyyy		New
-01	dd/mm/yyyy		Added..................
-02	dd/mm/yyyy		Modified...............

APPENDIX VI

TEMPLATE FOR AN EXAMPLE METHODS VALIDATION PROTOCOL

Methods Validation Protocol for Assay and Impurities for

Compound A in Your Product Tablets, 5- and 10-mg,

by High-Performance Liquid Chromatography

Protocol Number: P1234

Associated Development Report Number DR1234

PROTOCOL APPROVAL

APPROVED BY :

_____   _____

First, Last   Date

Scientist

_____   _____

First, Last   Date

Front Line Manager

_____   _____

First, Last   Date

Second Level Manager

_____   _____

First, Last   Date

Quality Assurance Representative

Issue Date: DD/MMM/YYYY

I. STUDY

This protocol was generated and approved to validate a high-performance liquid chromatographic (HPLC) stability indicating method for the analysis of compound A and its impurities related A and related B in your product 5- and 10-mg tablets.

The validation will be conducted in accordance with the *United States Pharmacopoeia*, International Conference of Harmonization guidelines and your SOP, "Validation of Analytical Test Procedures," SOP Number ABC-1243 rev4 dated 1 April 2002.

The experimental method was developed by your company's analytical development department and reported in development report DR1234, "Method Development Report for the Analysis of Compound A and Related Compounds A and B in Drug Product," and is included in Attachment A.

The following analytical performance characteristics or validation tests will be determined during this validation:

1. Accuracy (recovery)
   - Wet spiked placebo
   - Dry spiked placebo
2. Precision
   - Repeatability (method precision)
   - Intermediate precision
3. Specificity and/or selectivity
4. Detection limit
5. Quantitation limit
   - Relative response factor determination
6. Linearity
7. Range
8. Robustness testing
9. System suitability determination
10. Forced degradation
11. Solution stability
12. Filter retention
13. Extraction efficiency

II. MATERIALS

The following materials will be used during the execution of this protocol:

Name	Source	Description
Compound A , 5-mg tablets, lot 2, aged	Yours	Product
Compound A , 10-mg tablets, lot 3, aged	Yours	Product
Compound A , 5-mg tablets, lot 4	Yours	Product
Compound A , 10-mg tablets, lot 5	Yours	Product
Compound A placebo, (5.0 mg), lot 6	Yours	Product
Compound A placebo, (10.0 mg), lot 7	Yours	Placebo
Compound A , micronized, lot 1, purity 100.2%	Yours	API
Compound A reference standard, USP Lot K 100.0% Standard	USP	API reference
Related A lot 200109240001, purity 98.6% Standard	Yours	Impurity reference
Related B lot RS990002711, purity 99.25% reference standard	Yours	Impurity/degradant
FD & C Red #40 AL-LAKE HT, lot Z	Yours	Excipient

Compound A Tablet, 5 mg/Unit Formulation

#	mg/Unit	Ingredient Name
1	5.0	Compound A
2	25.0	Microcrystalline cellulose NF (Avicel PH 101)
3	1.5	Docusate sodium USP/sodium benzoate NF 80%/20%
4	3.0	Sodium starch glyconate NF
5	114.5	Lactose monohydrate NF (Fast Flo)
6	5.0	Stearic acid NF
7	0.5	Magnesium sterate NF
8	0.5	FD & C Red #40 AL-LAKE HT
	155	Total weight

Compound A Tablet 10 mg/Unit Formulation

#	mg/unit	Ingredient Name
1	10.0	Compound A
2	25.0	Microcrystalline cellulose NF (Avicel PH 101)
3	1.0	Docusate sodium USP/sodium benzoate NF 80%/20%
4	3.0	Sodium starch glyconate NF
5	110.5	Lactose monohydrate NF (Fast Flo)
6	5.0	Stearic acid NF
7	0.5	Magnesium sterate NF
	155	Total weight

The following specifications apply to compound A in both the 5- and 10-mg tablets:

Assay

NLT 90.0%, NMT 110.0%

Degradants and Impurities

Total impurities: NMT 1.0%

Related A: NMT 0.5%

Related B: NMT 0.5%

Unknown impurities: NMT 0.25%

The active pharmaceutical ingredient (API) contained in your product 5- and 10-mg tablets is compound A and possesses the following molecular structure and weight:

$C_{13}H_{17}N$ MW 187.29

Compound A API is manufactured by the following synthetic process:

Compound A possesses two potential impurities: related A and related B. Related A is a process impurity and related B is both a process impurity and potential degradant. Molecular structures and weights of both are given as follows.

Related A possesses the following molecular structure and weight:

$$C_{10}H_{15}N \quad MW \ 149.23$$

Related B possesses the following molecular structure:

$$C_{11}H_{13}N \quad MW \ 159.23$$

Degradation occurs through exposure to light via the following proposed mechanism:

Compound A   Related A   Related B

III. SAFETY PRECAUTIONS

Read the material safety data sheets for compound A, related A and B (if available), and all compounds under the reference material section of this report. Take all necessary precautions when using each compound. Laboratory safety wear will include a lab coat and safety glasses. Samples and test solutions containing these compounds will be handled, stored, and disposed in accordance with applicable (your company) standard operating procedures, and all applicable state and federal regulations.

IV. METHODS VALIDATION ANALYTICAL PERFORMANCE CHARACTERISTICS TO BE EVALUATED

NOTE: In addition to the instructions listed here, detailed sample preparation steps are shown in Attachment B: Sample Preparation Diagrams for all samples required during testing.

1. Accuracy (Recovery)

Description of Accuracy (Recovery)

Accuracy expresses the closeness of agreement between the value found and the value that is accepted as either a conventional true value or an accepted reference value. It may often be expressed as the recovery by the assay of known, added amounts of analyte. Samples (spiked placebos) are prepared normally covering 50% to 150% of the nominal sample preparation concentration. These samples are analyzed and the recoveries of each are calculated. Spiking can be performed as wet (e.g., via solution) or dry.

Experimental Determination of Accuracy (Recovery) for Assay

Accuracy is performed on the 5-mg tablets only since the 10-mg strength is a scale-up of the 5-mg strength and the 5 mg represents the worst case scenario with respect to excipients and potential recovery interferences. Prepare 15 total samples by weighing and spiking placebo as described in the sample preparation diagrams (the placebo remains 100% of method concentration in all samples).

- Spike the placebo samples with one stock active solution at each of the five levels and QS with diluent as per the method.
- Prepare three (3) replicate weights for each level.
- Prepare three samples at about 50%.
- Prepare three samples at about 70%.
- Prepare three samples at about 100%.
- Prepare three samples at about 130%.
- Prepare three samples at about 150%.
- Also prepare 3 dry spiked placebo samples at the nominal 100% tablet concentration for both the 5-mg and 10-mg dosage strengths. This is accomplished by accurately weighing compound A into placebo and mixing the dry aliquot into the placebo mixture.
- Inject each sample three times and analyze according to the analytical method, adequately bracketed by standard.
- Inject samples from the lowest concentration to the highest concentration.
- Calculate the % RSD for each individual weight at each level.
- Plot the analyte concentration for each individual weight versus the signal response (average of each set of injections).
- Perform linear regression analysis, but do not include the origin as a point made and do not force the line through the origin.
- Plot the sign and magnitude of the residuals versus analyte concentration.

- □ Check residual plot for outlying values and curvature.
- □ Evaluate *y* intercept to determine if there is significant departure from zero.
- □ Calculate the recovery for each individual sample weight (average of the three injections).
- □ Calculate the average recovery of the three sample weights at each concentration level.

Acceptance Criteria:

- □ The percent recovery of the spiked placebos should be within 100 ±2.0% for the average of each set of three weights.
- □ Each individual sample recovery should lie within the range of 98% to 102%.
- □ Coefficient of determination (r^2) should be greater than 0.9998.
- □ There should be no curvature in the residuals plot.
- □ The *y* intercept should not significantly depart from zero (e.g., area response of *y* intercept should be less than 5% of the response of the nominal 100% concentration value).

Experimental Determination of Accuracy (Recovery)
for Related A and B

Accuracy was performed on the 5-mg tablets only, since the 10-mg strength is a scale-up of the 5-mg strength and the 5 mg represents the worst case scenario with respect to excipients and potential recovery interferences. Assume that LOQ is 0.05%.

- □ Prepare 15 samples by weighing an appropriate amount of placebo (the placebo remains 100% of method concentration in all samples) with respect to the concentration specified in the method being validated.
- □ Spike each of the 15 placebo samples with active solution at 100% of the nominal active concentration.
- □ Prepare spiking solutions for each of the impurities.
- □ Spike the 15 placebo samples with each impurity in the following fashion:
- □ Prepare three replicate samples at about the LOQ concentration.
  - □ Prepare three replicate samples at about 25% of the impurity/degradant limit for each impurity.
  - □ Prepare three replicate samples at about 50% of the impurity/degradant limit for each impurity.

- □ Prepare three replicate samples at about 100% of the impurity/degradant limit for each impurity.
- □ Prepare three replicated samples at about 150% of the impurity-/degradant limit for each impurity.
- □ Inject each sample three times and analyze according to the analytical method, adequately bracketed by standard.
- □ Inject samples from the lowest concentration to the highest concentration.
- □ Calculate the % RSD for each individual weight at each level.
- □ Plot the analyte concentration for each individual weight versus the signal response (average of each set of injections).
- □ Perform linear regression analysis, but do not include the origin as a point made and do not force the line through the origin.
- □ Plot the sign and magnitude of the residuals versus analyte concentration.
- □ Check residual plot for outlying values and curvature.
- □ Evaluate y intercept to determine if there is significant departure from zero.
- □ Calculate the recovery for each individual sample weight (average of the three injections).
- □ Calculate the average recovery of the three sample weights at each concentration level.

Acceptance Criteria:

- □ The individual sample recovery of each impurity should be within 75% to 125%.
- □ The average percent recovery for each impurity should lie within the range of 75% to 125%.
- □ Coefficient of determination (r^2) should be greater than 0.995.
- □ There should be no curvature in the residuals plot.
- □ The y intercept should not significantly depart from zero (e.g., area response of y intercept should be less than 5% of the response of the nominal 100% concentration value).

Notes   LOD and LOQ experiments should be conducted prior to the accuracy experiments.

2. Precision

a. Repeatability (Method Precision)

Description of Repeatability (Method Precision)

Repeatability evaluates the variation experienced by a single analyst on a single instrument. Repeatability does not distinguish between variation from the instrument or system alone and from the sample preparation process. Repeatability is performed by analyzing multiple replicates of an assay composite sample using the analytical method. The recovery value is calculated and reported for each value.

Experimental Determination of Repeatability (Method Precision) for Assay

- □ Prepare six replicate samples solutions, for both the 5-mg and 10-mg tablet strength from the same assay composite sample according to the analytical method.
- □ Analyze the samples according to the analytical method and make two injections of each sample.
- □ Calculate the assay results (% recovery) for each sample.
- □ Calculate the mean and relative standard deviations (% RSD) of the six sample preparations.

Acceptance Criteria:

- □ The % RSD of the assay or recovery values should not be greater than 2.0%.

b. Intermediate Precision

Description of Intermediate Precision

Intermediate precision refers to variations within a laboratory as with different days, with different instruments, by different analysts, and so forth. Intermediate precision was formally known as ruggedness. A second analyst repeats the repeatability analysis on a different day using different conditions and different instruments. The recovery values are calculated and reported. A statistical comparison is made to the first analyst's results.

Experimental Determination of Intermediate Precision for Assay

With the following restrictions, have a second analyst execute the following steps:

- □ Perform the repeatability analysis on different days.
- □ Perform the repeatability analysis using different operating conditions and different instruments when possible (e.g., column, apparatus, reagents)
- □ If possible use a different manufacturer's instrument.
- □ Prepare the six replicate samples solutions from the same assay composite sample according to the analytical method.
- □ Analyze the samples according to the analytical method make two injections of each sample.
- □ Calculate the assay results (% recovery) for each sample.

Acceptance Criteria:

- □ The % RSD of the assay/recovery values generated by a single analyst should not be greater than 2.0%.
- □ The % RSD of the combined assay/recovery values generated by both analysts, over both days should not be greater than 3.0%.

Notes   A chromatogram of a typical standard, a typical system suitability standard, and a typical sample should be included in the validation report.

Experimental Determination of Repeatability (Method Precision) for Related A and B

- □ Prepare six replicate sample solutions from the same assay composite sample according to the analytical method.
- □ If the impurities are not present in significant amounts ($>0.25\%$), spike each sample with impurities at the limit of 0.50%.
- □ Analyze the samples according to the analytical method and make two injections of each sample.
- □ Calculate the assay results (% recovery) for each sample.
- □ Calculate the individual impurities results (% recovery) for each sample.

□ Calculate the mean and relative standard deviations (% RSD) of the six sample preparations for assay and impurities.

Acceptance Criteria:

□ The % RSD of the related compounds recovery values should not be greater than 15%.

Experimental Determination of Intermediate Precision for Related A and B

With the following restrictions, have a second analyst execute the following steps:

□ Perform the repeatability analysis on different days.
□ Perform the work using different operating conditions and different instruments when possible (e.g., column, apparatus, reagents).
□ If possible use a different manufacturer's instrument.
□ Prepare six replicate sample solutions from the same assay composite sample according to the analytical method.
□ If the impurities are not present in significant amounts ($>0.25\%$), spike each sample with impurities at the limit of 0.50%.
□ Analyze the samples according to the analytical method and make two injections of each sample.
□ Calculate the assay results (% recovery) for each sample.
□ Calculate the individual impurities results (% recovery) for each sample.
□ Calculate the average total percent impurities/degradants in each sample or recovery values and the % RSD of all the individual sample weights over both days.

Acceptance Criteria:

□ The % RSD of the impurities/degradants generated on the second day should not be greater than 15%.
□ The % RSD of the combined assay/recovery values generated by both over both days should not be greater than 15%.

Notes   A chromatogram of an impurity typical standard, a typical system suitability standard, a typical sample, and a sample spiked with impurities should be included in the validation report.

In cases where the impurity quantities are limited, the same stock impurity solution can be used by both analysts.

3. Specificity and/or Selectivity

Description of Specificity and/or Selectivity

Specificity is the ability to assess unequivocally the analyte in the presence of components that may be expected to be present such as impurities, degradation products, and excipients. There must be inarguable data for a method to be specific. Specificity measures only the desired component without interference from other species which might be present; separation is not necessarily required. Selectivity are the ability of the analytical method to resolve each and every related compound in the mixture. Specificity is required for assay but selectivity is not. Both specificity and selectivity are required for impurities analysis. Specificity and selectivity are determined by analyzing blanks, sample matrix (placebo), and known related impurities to determine whether interferences occur. Specificity and selectivity are also demonstrated during forced degradation studies.

Experimental Determination of Specificity and/or Selectivity

- Make an injection of a blank/diluent during each chromatographic run to ensure there are no interferences.
- Prepare a placebo for both the 5-mg and 10-mg tablet strength.
- Inject each placebo preparation twice.
- Confirm that no peaks can be attributed to the blank/diluent or placebo.
- Define any peaks observed by RRT indexed to the active component.
- Prepare and inject twice, samples of related A and related B at the impurity/degradant (e.g., related compound) specification limit of 0.50%.
- Prepare two separate spiked solutions containing the active at 100% and each impurity (from two separate impurity preparations) at the limit.
- Inject the spiked samples twice to confirm specificity.

Acceptance Criteria:

- Confirm that no interference in the elution zone of the active occurred from the blank/diluent, internal standard (if applicable), or the placebo.
- Confirm that the impurities/degradants do not elute in the elution zone of the active.

Notes   Primary reference standards should be used if at all possible.

4. Detection Limit

Description of Detection Limit

The detection limit (DL) or limit of detection (LOD) of an individual procedure is the lowest amount of analyte in a sample that can be detected but not
necessarily quantitated as an exact value. In analytical procedures that exhibit
baseline noise, the LOD can be based on a signal-to-noise ratio (3:1), which is
usually expressed as the concentration (e.g., percentage, parts per billion) of
analyte in the sample. There are several ways in which it can be determined,
but usually involves injecting samples which generate a S/N ratio of 3:1 and
estimating the DL.

Experimental Determination of Detection Limit for Assay

Not performed for Category I (assay) methods.

Experimental Determination of Detection Limit for Related A and B

- Prepare standard solution and stock solution for each impurity/degradant
  and the active.
- Prepare a combined sample created by mixing the impurities and active
  into one solution.
- Perform serial dilutions to create solutions that span from 150% of the
  impurity/degradant specification to 0.005%. Levels should include:
  - 0.005%
  - 0.010%
  - 0.025%
  - 0.050%
  - 0.10%
  - 0.25%
  - 0.50% (specification limit)
  - 0.75%
  Note: This test may be done for each compound individually if necessary.
- Make five injections of each concentration, adequately bracketed by the
  standard.

- Inject sample solutions from the lowest concentration to the highest concentration.
- Calculate the % RSDs at each concentration.
- Calculate the signal-to-noise ratio for each peak in each sample at each level.
- The limit of detection is the first concentration at which the analyte has a signal-to-noise ratio of 3:1.

Acceptance Criteria:

- The limit of detection is the first concentration at which the analyte has a signal-to-noise ratio of 3:1.

Notes   The purest analyte available should be used to determine the limit of detection and limit of quantitation. (i.e., primary standard such as USP or EPCRS).

5.   Quantitation Limit

Description of Quantitation Limit

The quantitation limit (QL) or limit of quantitation (LOQ) of an individual analytical procedure is the lowest amount of analyte in a sample that can be quantitatively determined with suitable precision and accuracy. The quantitation limit is a parameter of quantitative assays for low concentrations of compounds in sample matrices and is used particularly for the determination of impurities and/or degradation products. It is usually expressed as the concentration (e.g., percentage, parts per million) of analyte in the sample. For analytical procedures that exhibit baseline noise the LOQ is generally estimated from a determination of signal-to-noise ratio (10:1) and is usually confirmed by injecting standards that give this S/N ratio and have acceptable % RSDs as well.

Experimental Determination of Quantitation Limit for Assay

Not performed for Category I (assay) methods.

Experimental Determination of Quantitation Limit for Related A and B

- The quantitation limit is determined concurrently with the detection limit.
- The limit of quantitation is the level at which a signal-to-noise ratio of 10:1 is obtained and the difference between injections is less than 10%.

Acceptance Criteria:

The limit of quantitation is the level at which a signal-to-noise ratio of 10:1 is obtained and the difference between injections is less than 10%.

Experimental Determination of Relative Response Factors (RRF) for Related A and B

☐ Plot the analyte concentration for each set of dilutions separately versus the signal response (average of each set of injections) for the data obtained during the LOD/LOQ experiments.

☐ Perform linear regression analysis, but do not include the origin as a point made and do not force the line through the origin.

☐ Plot the sign and magnitude of the residuals versus analyte concentration.

☐ Check residual plot for outlying values and curvature.

☐ Evaluate *y* intercept to determine if there is significant departure from zero.

☐ The relative response factors for related A and B are calculated by dividing the slope of compound A by the slope obtained for related A and B.

6. Linearity

Description of Linearity

Linearity evaluates the analytical procedure's ability (within a given range) to obtain a response that is directly proportional to the concentration (amount) of analyte standard. If the method is linear, the test results are directly, or by well-defined mathematical transformation, proportional to the concentration of analyte in samples within a given range. Note that this is different from *range* (sometimes referred to as *linearity of method*), which is evaluated using samples and must encompass the specification range of the component assayed in the drug product. Linearity may be established for all active substances, preservatives, and expected impurities. Evaluation is performed on standards.

Experimental Determination of Linearity for Assay

☐ Prepare five standard solutions of the analyte at ~50%, 70%, 100%, 130%, and 150% of the method concentration using serial dilutions from a stock solution.

☐ Make three injections at each concentration, adequately bracketed by the standard.

☐ Make sure to inject samples from the lowest concentration to the highest concentration to reduce the effects, if any, of carryover from the higher concentration samples.

- □ Calculate the % RSD at each concentration.
- □ Plot the analyte concentration for each set of dilutions separately versus the signal response (average of each set of injections).
- □ Perform linear regression analysis, but do not include the origin as a point made and do not force the line through the origin.
- □ Plot the sign and magnitude of the residuals versus analyte concentration.
- □ Check residual plot for outlying values and curvature.
- □ Evaluate y intercept to determine if there is significant departure from zero.

Acceptance Criteria:

- □ Coefficient of determination (r^2) should be greater than 0.9999.
- □ There should be no curvature in the residuals plot.
- □ The y intercept should not significantly depart from zero (e.g., area response of y intercept should be less than 5% of the response of the midrange concentration value).

Experimental Determination of Linearity for Related A and B

The results of the detection limit and quantitation limit determinations are to be used for this experiment.

7. Range

Description of Range

Range is the interval between the upper and lower concentrations (amounts) of analyte in the sample (including these concentrations) for which it has been demonstrated that the analytical procedure has a suitable level of precision, accuracy, and linearity. Range is normally expressed in the same units as test results (e.g., percent, parts per million) obtained by the analytical method. Range (sometimes referred to as *linearity of method*) is evaluated using samples (usually spiked placebos) and must encompass the specification range of the component assayed in the drug product.

Experimental Determination of Range for Assay and Related A and B

Use data obtained from accuracy (recovery) experiments in Section 1.

Acceptance Criteria:

- □ Coefficient of determination (r^2) should be greater than 0.9998.
- □ There should be no curvature in the residuals plot.

▫ The *y* intercept should not significantly depart from zero (e.g., area response of *y* intercept should be less than 5% of the response of the nominal 100% concentration value).

8. Robustness Testing

Description of Robustness Testing

Robustness is the measure of the ability of an analytical method to remain unaffected by small but deliberate variations in method parameters (e.g., pH, mobile phase composition, temperature, instrument settings) and provides an indication of its reliability during normal usage. Robustness testing is a systematic process of varying a parameter and measuring the effect on the method by monitoring system suitability and/or the analysis of samples.

Experimental Determination of Robustness for Assay and Related A and B

▫ Prepare an assay standard as described in the analytical method.
▫ Prepare an impurities standard as described in the analytical method.
▫ Prepare standard solution containing active at the 100% level spiked with known impurities/degradants at the limit of 0.50%.

Mobile Phase Variation Using the Mobile Phase Specified in the Test Method:

▫ Increase each [major] component by 5%, inject the system suitability standard twice, and measure the appropriate figures of merit.
▫ Decrease each [major] component by 5%, inject the system suitability standard twice, and measure the appropriate figures of merit.
▫ Increase each [major] component by 10%, inject the system suitability standard twice, and measure the appropriate figures of merit.
▫ Decrease each [major] component by 10%, inject the system suitability standard twice, and measure the appropriate figures of merit.
▫ Increase each [minor] component by 15%, inject the system suitability standard twice, and measure the appropriate figures of merit. (*Note:* Minor components are those with less than 10 mL/L.)
▫ Decrease each [minor] component by 15%, inject the system suitability standard twice, and measure the appropriate figures of merit.
▫ Increase each [minor] component by 30%, inject the system suitability standard twice, and measure the appropriate figures of merit.

□ Decrease each [minor] component by 30%, inject the system suitability standard twice, and measure the appropriate figures of merit.

HPLC Column Temperature Variation Using a Column Heater:

□ Inject the system suitability standard injected twice at the temperature stated in the method, and measure the appropriate figures of merit.

□ Inject the system suitability standard injected twice at +5° above stated method temperature and measure the appropriate figures of merit.

□ Inject the system suitability standard injected twice at −5° above stated method temperature, and measure the appropriate figures of merit.

Mobile Phase Flow-Rate Variation:

□ Inject the system suitability standard twice at a 10% increase in flow rate, and measure the appropriate figures of merit.

□ Inject the system suitability standard injected twice at a 10% decrease in flow rate, and measure the appropriate figures of merit.

□ Inject the system suitability standard injected twice at a 25% increase in flow rate, and measure the appropriate figures of merit.

□ Inject the system suitability standard injected twice at a 25% decrease in flow rate, and measure the appropriate figures of merit.

For Buffer pH Variation:

□ Increase the mobile phase 0.25 pH units, inject system suitability standard injected twice, and measure the appropriate figures of merit.

□ Decrease the mobile phase 0.25 pH units, inject system suitability standard injected twice, and measure the appropriate figures of merit.

HPLC Column Variation:

□ Using three columns from at least two different lots of packing material obtained, inject the system suitability standard injected twice on each column, and measure the appropriate figures of merit.

□ Using a brand new column and an old column (>500 injections), inject system suitability standard injected twice on each column and measure the appropriate figures of merit.

Acceptance Criteria:

□ Measure appropriate figures of merit (e.g., resolution, tailing factor, theoretical plates, and capacity factor) for each variation experiment.

□ Determine the suitability of the method under each modification determined by taking into account peak shape, retention time, system pressure, and system suitability parameters.

- Determine which system suitability parameters are important to the overall function of the method.
- Establish limits for critical parameters.
- For column variability ensure all columns used in validation are commercially available.
- Ensure that three columns from at least two different lots of packing material are obtained and used.
- Ensure that a brand new column and an old column (> 500 injections) are obtained and used.
- Ensure the retention times are similar on each of the three columns.

9. System Suitability Determination

Description of System Suitability Testing

System suitability is the evaluation of the components of an analytical system to show that the performance of a system meets the standards required by a method. A system suitability evaluation usually contains its own set of parameters; for chromatographic assays, these may include tailing factors, resolution, and precision of standard peak areas, and comparison to a confirmation standard, capacity factors, retention times, theoretical plates, and calibration curve linearity. Where applicable, system suitability parameters are calculated, recorded, and trended throughout the course of the validation. Final values are then determined from this history.

Experimental Determination of System Suitability for Assay and Related A and B

- Where applicable, calculate system suitability (capacity factor, tailing factor, resolution, theoretical plates, and reproducibility) for each chromatographic run during methods validation.
- Identify upper and lower limits for each parameter by analyzing values obtained throughout the validation.
- Establish minimum criteria from this range.

Acceptance Criteria:

The following minimum criteria are suggested:

- $k' \geq 2.0$, where k' is capacity factor
- $T \leq 2$, where T is tailing factor
- $R > 1.5$, where R is resolution

- $N \geq 1000$ plates, where N is theoretical plates per column
- % RSD $\leq 2.0\%$, where % RSD is the percent relative standard deviation

10. Forced Degradation

Description of Forced Degradation

Forced degradation studies are undertaken to degrade the sample (e.g., drug product or API) deliberately. These studies are used to evaluate an analytical method's ability to measure an active ingredient and its degradation products without interference. Samples or drug product (spiked placebos) and drug substance are exposed to heat, light, acid, base, and oxidizing agent to produce 10%–30% degradation of the active. The degraded samples are then analyzed using the method to determine if there are interferences with the active or related compound peaks.

Experimental Execution of Forced Degradation for Assay and Related A and B

The following degradation conditions are recommended as a starting point:

Degradation Reaction	Typical Conditions
Acid hydrolysis	1. Sample in aqueous acid or acidified solvent at ~0.5 N up to 24 hours (or)
	2. Heat/reflux or UV radiation in ~0.5 N HCl up to 24 hours.
Base hydrolysis	1. Sample in aqueous base or basic solvent at ~0.5 N up to 24 hours (or)
	2. Heat/reflux or UV radiation in ~0.5 N NaOH up to 24 hours.
Oxidation	1. Treat with ~3% H_2O_2 up to 24 hours (or)
	2. UV irradiation in ~3% H_2O_2 up to 24 hours.
Light decomposition (photolysis)	Expose to high-intensity UV light in suitable increments, up to 24 hours.
Thermal decomposition (pyrolysis)	Expose to ~100° C heat in suitable increments, up to 24 hours.

- Obtain or prepare solutions of ~0.5 N HCL, ~0.5 N NaOH, ~3% H_2O_2, and purified water.
- Prepare blank samples for each condition, including light and heat.

- ▫ Prepare a standard, spiked placebo, and unspiked placebo as appropriate in preparation for degradation.
- ▫ Analyze samples and blanks according to the method outlined in the validation prior to degradation, establishing an initial time point.
- ▫ Expose these standards, spiked placebos, and an unspiked placebo to acid, base, oxidation, heat, and light using the condition guidelines listed above.
- ▫ Neutralize acid and base samples prior to analysis by pipetting in an amount of acid or base solution equal to the sample aliquot, and then diluting to volume with water, diluent, or mobile phase.
- ▫ Analyze samples and blanks at varying intervals over 24 hours, according to the method outlined in the validation following degradation. Assess whether sufficient degradation (10%–30%) has occurred.
- ▫ Calculate the percent recovery of the appropriate solutions to determine the extent of degradation.
- ▫ If the acid, base, and peroxide solutions were not sufficiently degraded in 24 hours, the acid, base, and peroxide solutions should have been exposed to heat/light, until at least 10%–30% degradation is achieved or 24 hours of exposure has elapsed. Moreover, if the samples are overdegraded, lessening the length and severity of the conditions to obtain the 10%–30% is acceptable.
- ▫ Perform peak purity analysis on the main analyte peak using diode array and/or mass spectra analysis.
- ▫ Capture and display chromatograms for each degradation on appropriately degraded (10%–30%) samples.
- ▫ Capture and display peak purity analyses for each degradation condition.
- ▫ Determine whether the method is specific for the degraded samples.

Acceptance Criteria:

- ▫ Confirm that peak purity of the degraded spiked placebo sample main peak was performed using UV or MS analysis.
- ▫ Ensure peak purity was performed for each degradation condition and that the samples have degraded sufficiently or degradation had stopped.
- ▫ Evaluate spectral overlay of the sufficiently degraded impurities/degradants spiked placebos in addition to peak purity to demonstrate that the degradants are resolved from the analyte.
- ▫ Determine relative retention times of the degradants.
- ▫ Assess chromatograms of the sufficiently degraded spiked placebo overlaid with the degraded placebo under each degradation condition.

Notes Forced degradation studies are designed to produce potential degradation products that may be encountered in real-world scenarios. Degradants generated may or may not be what is seen during stability studies.

Performing the actual degradation in these studies is not an exact science and may require modifying conditions to obtain the desired 10%–30% degradation. However, if the maximum conditions listed here do not produce degradation, then it is not necessary to continue the experiments until degradation occurs. A statement is then added to the report confirming the stable nature of the molecules.

Targeted recovery for the initial samples prior to degradation should be in the 95%–105% range.

The run time for forced degradation samples should be sufficiently long to observe the retention time of the latest eluting active or degradant.

11. Solution Stability

Description of Solution Stability

The stability of standards and samples is established under normal benchtop conditions, normal storage conditions, and sometimes in the instrument (e.g., an HPLC autosampler) to determine if special storage conditions are necessary, for instance, refrigeration or protection from light. Stability is determined by comparing the response and impurity profile from aged standards or samples to that of a freshly prepared standard and to its own response from earlier time points.

Experimental Determination of Solution Stability

- Prepare fresh standard and spiked placebo (or actual tablet/capsule) as per the test method. Ensure that both stock and working solutions are available for analysis.
- Analyze these solutions analyzed as per the test method, establishing a time zero value for each.
- Place an aliquot of each solution in clear glassware and expose to ambient (benchtop) conditions.
- Place an aliquot of each solution placed in amber glassware and expose to ambient (benchtop) conditions.
- Place an aliquot of each solution in clear glassware and place in a refrigerator.

□ Analyze these samples versus fresh standard every 24 hours for at least two (2) days.

□ Make two injections made of each solution.

□ Calculate the percent recoveries calculated for all solutions.

Acceptance Criteria:

□ For assay level standards, the fresh standard and the verification standard should not differ more that 1.0%.

□ For the assay level, the standard and sample solutions are considered sufficiently stable over time if the recovery value does not vary more than ±1.5% (absolute) from the initial result.

Notes   After two days, samples may be evaluated at intervals at the analyst's discretion. A fresh standard and fresh verification standard should be prepared each day.

12.   Filter Retention

Description of Filter Retention

Filter retention studies are a comparison of filtered to unfiltered solutions during a methods validation to determine whether the filter being used retains any active compounds or contributes unknown compounds to the analysis. Blank, sample, and standard solutions are analyzed with and without filtration. Comparisons are made in recovery and appearance of chromatograms.

Experimental Determination of Filter Retention for Assay and Related A and B

□ Prepare three placebos for the lowest potency, spiked with active at 100% of the nominal assay concentration, and each of the impurities at the limits.

□ Where necessary, three product sample preparations (tablets, capsules, drug substance, etc.) containing the known impurities in sufficient quantities (~0.5%) may be substituted for the spiked placebos.

□ Filter portions of each individual spiked placebo sample filtered through at least two (2) prospective filters. Ensure the filters are commercially available. Centrifuge an aliquot of each individual spiked placebo (e.g., 3000 rpm for 10 minutes suggested).

- Inject all three samples three times each, and analyze according to the analytical method adequately bracketed by the standard.
- Calculate the recovery of active for each individual sample weight and % RSD of the replicate injections.
- Calculate the recovery of each individual impurity for each sample weight and % RSD of the replicate injections.
- Calculate the average recovery for each filter.
- Calculate the percent difference of the average result for each filter versus the average result of the centrifuged sample.

Acceptance Criteria:

- For the spiked placebos, the percentage recovery should be $100 \pm 2.0\%$ for the average of each set of three weights for assay.
- For each individual spiked placebo the recovery should be 98%–102% for assay.
- For an acceptable filter, the difference between the filtered sample and the centrifuged sample should be NMT 1.5% (absolute).
- For an acceptable filter, the difference between the filtered sample and the centrifuged sample for impurities is NMT 5% (absolute).
- The individual and average recovery of each impurity should be within the range of 75% to 125%.

13. Extraction Efficiency

Description of Extraction Efficiency

Extraction efficiency is the measure of the effectiveness of extraction of the drug substance from the sample matrix. Studies are conducted during methods validation to determine that the sample preparation scheme is sufficient to ensure complete extraction without being unnecessarily excessive. Extraction efficiency is normally investigated by varying the shaking or sonication times (and/or temperature) as appropriate during sample preparation on manufactured (actual) drug product dosage forms.

Experimental Determination of Extraction Efficiency for Assay

- Using aged samples, preferably an expired lot, obtain twelve sample weights for both 5-mg and 10-mg tablet strengths.
- Prepare the first set of three samples prepared as per the sample procedure in the method (10 minutes of sonication).
- Prepare the second set of three samples prepared and extracted longer than the first set.

▫ Prepare the third set of three samples prepared and extracted longer than the second set.
▫ Prepare the fourth set of three samples prepared and extracted for a shorter period of time than that specified extraction time in the procedure.
▫ Calculate the results in percent recovered.

Acceptance Criteria:

▫ Establish an extraction range for the sample preparation procedure after evaluation of the data.

Notes It should be demonstrated that the extraction of the drug from the sample matrix is sufficient to ensure complete extraction without being unnecessarily excessive.

V. REPORT

Upon completion of the methods validation experiments, a draft report will be presented to your laboratory for, evaluation, review, and comment, after which a final report will be issued. Your company's development quality assurance department will audit final reports.

VI. MAINTENANCE OF RAW DATA

Original data, or copies, will be available at your company's development to facilitate auditing the study during its progress, and before the issuance of a final report. When the final report is completed, all original paper data, all magnetically encoded records, and a copy of the final report will be retained in the archives of your company's development department.

VII. REFERENCES

Your SOP, "Validation of Analytical Test Procedures," SOP Number ABC-1243 rev4 dated 1 April 2002.

VIII. REVISION SUMMARY

Original version.

ATTACHMENT A—ANALYTICAL METHOD

I. Reagents

Sodium acetate, HPLC grade

Glacial acetic acid, HPLC grade

THF, HPLC grade

Acetonitrile, HPLC grade

Water, in-house deionized water

Compound A reference standard

II. Mobile Phase Preparation

Mobile phase A: 70:24:6 of 25 mM sodium acetate buffer (pH 5.0)

Acetonitrile: THF

Mobile phase B: 60:40 THF:acetonitrile

Preparation of 25 mM sodium acetate buffer, pH 5.0

Accurately weigh 9.6 g of sodium acetate and quantitatively transfer to a 5L container. Add 5 L of purified water and mix until dissolved. Adjust the pH to 5.0 ($\pm$0.05) using glacial acetic acid. This solution preparation may be scaled up as necessary.

Preparation of mobile phase A (70:24:6::sodium acetate buffer (pH 5.0):acetonitrile:THF (v:v:v))

Prepare mobile phase A to have a ratio of 70:24:6 sodium acetate buffer (pH 5.0):acetonitrile:THF. For example, to prepare 4 L, combine 2800 mL of sodium acetate buffer (pH 5.0), 960 mL of acetonitrile, and 240 mL of THF. Mix well and degas by sonication prior to use.

Preparation of mobile phase B (60:40::THF:acetonitrile)

Prepare mobile phase B to have a ratio of 60:40 THF:acetonitrile. For example, to prepare 4 L, combine 2400 mL of THF and 1600 mL of acetonitrile. Mix well and degas by sonication prior to use.

Diluent (50:50 mobile phase A:THF)

Prepare a solution of mobile phase A and THF with a ratio of 50:50. For example, to prepare 4 L, combine 2000 mL of mobile phase A and 2000 mL of THF. Mix well.

III. Standard Preparation

Assay standard solution (0.1 mg/mL or 100 µg/mL) (prepare in duplicate)

Accurately weigh approximately 10 mg of compound A reference standard into a 100-mL volumetric flask. Add ~50 mL of diluent and sonicate for 10 minutes with intermittent shaking until a clear solution is obtained. Allow solution to cool to room temperature and dilute to volume with diluent. Label as main and check assay standards.

Impurity standard solution (0.0001 mg/mL or 0.1 µg/mL) (prepare in duplicate)

Transfer 10.0 mL of the assay standard solution into a 1000-mL volumetric flask. Dilute to volume with diluent and mix well. Transfer 10.0 mL of this solution into a 100-mL volumetric flask. Dilute to volume with diluent and mix well. Label as main and check impurity standards. This is 0.1% of the nominal sample concentration.

Sensitivity solution (0.00005 mg/mL or 0.05 µg/mL)

Transfer 10.0 mL of the assay standard solution into a 1000-mL volumetric flask. Dilute to volume with diluent and mix well. Transfer 5.0 mL of this solution into a 100-mL volumetric flask. Dilute to volume with diluent and mix well. This is 0.05% of the nominal sample concentration.

IV. Sample Preparation

Assay and impurities sample preparation (prepare in duplicate)

Accurately weigh 50 tablets, grind to fine powder, and calculate the average tablet weight. Accurately weigh duplicate samples of ~155 mg (for 5-mg tablets) and 77.5mg (for 10-mg tablets) of powder (equivalent to 5 mg of compound A) into separate 50-mL volumetric flasks. Fill approximately three by four full with diluent and sonicate for 10 minutes with intermittent shaking. Dilute to volume with diluent and centrifuge or filter a portion of the solution to obtain a clear solution. Nominal concentration is 0.1 mg/mL or 100 µg/mL.

Drug substance (API) sample preparation

Accurately weigh ~10-mg API into a 100-mL volumetric flask. Add~50 mL of diluent and sonciate for 10 minutes with intermittent shaking until a clear solution is obtained. Allow solution to cool to room temperature, and dilute to volume with diluent.

V. Chromatographic Conditions

Column:	Serta, Sleeper C4, 3.5 mm, 4.6 × 150 mm
Column temp:	30 °C
Mobile phase A:	70:24:6::25 mM sodium acetate buffer (pH 5.0):acetonitrile:THF (v:v:v)
Mobile phase B:	60:40::THF:acetonitrile (v:v)
Flow rate:	1.50 mL/min
Detector:	UV at 254 nm

Injection volume:	50 mL	
Run time:	35 min	
Gradient profile:	(ramps are linear)	

Time (min)	% A	% B
0	100	0
4	100	0
26	51	49
27	100	0
35	100	0

Assay

Assay standards and sample preparations may be analyzed according to the following injection sequence:

Test Solution	# of Injections
Blank (diluent)	2
Main standard	5
Check standard	2
Assay sample 1	1
Assay sample 2	1
Assay sample X	1
Main standard	1

Note: Inject samples, with no more than 10 sample injections between bracketing standard solutions.

Impurities

Impurity standards and sample preparations may be analyzed according to the following injection sequence:

Test Solution	# of Injections
Blank (diluent)	2
Sensitivity solution	1
Main impurity standard	6
Check impurity standard	2
Impurity sample 1	1
Impurity sample 2	1
Impurity sample X	1
Main impurity standard	1

Note: Inject samples with no more than 10 sample injections between bracketing standard solutions.

Assay and Impurities

Standards and sample preparations may be analyzed according to the following injection sequence:

Test Solution	# of Injections
Blank (diluent)	2
Sensitivity solution	1
Check impurity standard	2
Main impurity standard	6
Check standard	2
Main standard	5
Main impurity standard	1
Sample 1	1
Sample 2	1
Sample X	1
Main standard	1
Main impurity standard	1

Note: Inject no more than 8 samples between bracketing standards.

VI. System Suitability

System suitability criteria will be established after completion of the validation. The following minimum criteria should be met for the assay standard.

- $k'≥5.0$, where k' is the capacity factor
- $T≤2$, where T is the tailing factor
- $R≥2.0$, where R is the resolution between adjacent peaks
- $N≥15000$, where N is the number of theoretical plates
- % RSD≤2.0, where RSD is the relative standard deviation of five replicate standard injections
- Main check standard agreement $± 2.0\%$

Requirements for the impurity standards are:

- % RSD≤10.0, where % RSD is the percent relative standard deviation of six replicate standard injections
- Main check standard agreement $± 10.0\%$, required only if run is for impurity testing only; not needed if assay and impurity testing run together
- S/N≥10 for the compound A peak in the sensitivity solution

VII. Calculations

Compound A assay for tablets:

$$\text{Compound A (mg/tablet)} = \frac{A_{sam}}{A_{std}} \times \frac{Wt_{std} \times PF}{100 \text{ mL}} \times \frac{50 \text{ mL}}{Wt_{sam}} \times Wt_{avg}$$

$$\text{Compound A (\%label)} = \frac{\text{Compound A (mg/tablet)}}{TS} \times 100$$

Compound A assay for API:

$$\text{Compound A (\%w/w)} = \frac{A_{sam}}{A_{std}} \times \frac{Wt_{std} \times PF}{100 \text{ mL}} \times \frac{100 \text{ mL}}{Wt_{sam}} \times 100$$

% Impurity for tablets:

$$\text{\% Impurity (\%w/w)} = \frac{A_{imp}}{A_{istd}} \times \frac{Wt_{std} \times PF}{Wt_{sam}} \times \frac{Wt_{avg}}{TS}$$

$$\times \frac{50 \text{ mL} \times 10 \text{ mL}}{100 \text{ mL} \times 1000 \text{ mL}} \times RRF \times 100$$

% Impurity for API:

$$\text{\% Impurity (\%w/w)} = \frac{A_{imp}}{A_{istd}} \times \frac{Wt_{std} \times PF}{Wt_{sam}}$$

$$\times \frac{100 \text{ mL} \times 10 \text{ mL}}{100 \text{ mL} \times 1000 \text{ mL}} \times RRF \times 100$$

where

A_{sam}	=	area response of compound A peak in the sample preparation
A_{imp}	=	area response of an impurity peak in the sample preparation
A_{std}	=	average area response of compound A peak in the standard preparation
A_{istd}	=	average area response of compound A peak in the impurity standard preparation
Wt_{std}	=	weight of standard, mg
Wt_{sam}	=	weight of powder taken in sample, mg

Wt_{avg} = average weight of tablets, mg
PF   = standard purity factor
TS   = tablet strength, mg
RRF   = relative response factor (RRF = 1 for unknown impurities)

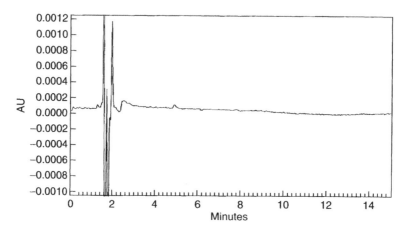

Example of blank injection.

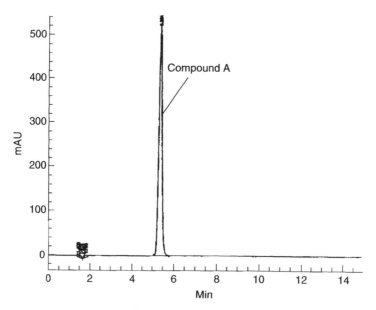

Example of standard injection.

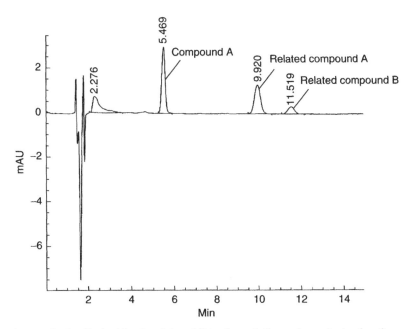

Impurity standard spiked with related A and B to show elution order and retention time of relateds.

Revision History

Version	Effective	Author	Summary of Change
-00	dd/mm/yyyy		New

ATTACHMENT B—SAMPLE PREPARATION DIAGRAMS

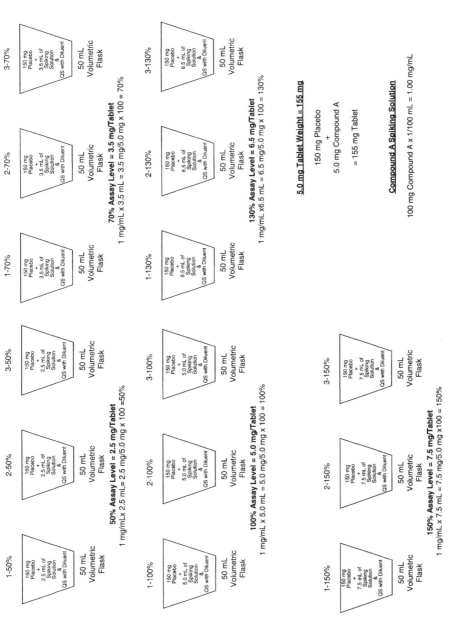

1-50% / **2-50%** / **3-50%**

150 mg Placebo + 2.5 mL of Spiking Solution & QS with Diluent — 50 mL Volumetric Flask

50% Assay Level = 2.5 mg/Tablet
1 mg/mL x 2.5 mL = 2.5 mg/5.0 mg x 100 =50%

1-70% / **2-70%** / **3-70%**

150 mg Placebo + 3.5 mL of Spiking Solution & QS with Diluent — 50 mL Volumetric Flask

70% Assay Level = 3.5 mg/Tablet
1 mg/mL x 3.5 mL = 3.5 mg/5.0 mg x 100 = 70%

1-100% / **2-100%** / **3-100%**

150 mg Placebo + 5.0 mL of Spiking Solution & QS with Diluent — 50 mL Volumetric Flask

100% Assay Level = 5.0 mg/Tablet
1 mg/mL x 5.0 mL = 5.0 mg/5.0 mg x 100 = 100%

1-130% / **2-130%** / **3-130%**

150 mg Placebo + 6.5 mL of Spiking Solution & QS with Diluent — 50 mL Volumetric Flask

130% Assay Level = 6.5 mg/Tablet
1 mg/mL x6.5 mL = 6.5 mg/5.0 mg x 100 = 130%

1-150% / **2-150%** / **3-150%**

150 mg Placebo + 7.5 mL of Spiking Solution & QS with Diluent — 50 mL Volumetric Flask

150% Assay Level = 7.5 mg/Tablet
1 mg/mL x 7.5 mL = 7.5 mg/5.0 mg x100 = 150%

5.0 mg Tablet Weight = 155 mg

150 mg Placebo
+
5.0 mg Compound A
= 155 mg Tablet

Compound A Spiking Solution

100 mg Compound A x 1/100 mL = 1.00 mg/mL

5-mg compound A tablet accuracy (recovery) sample preparation table.

203

ATTACHMENT B—SAMPLE PREPARATION DIAGRAMS

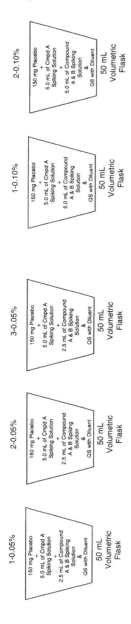

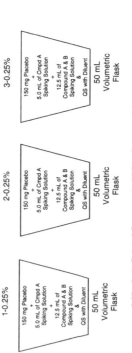

1-0.05% / 2-0.05% / 3-0.05%

Flasks (each): 150 mg Placebo + 5.0 mL of Cmpd A Spiking Solution + 2.5 mL of Compound A & B Spiking Solution & QS with Diluent — 50 mL Volumetric Flask

0.050 w/w % Related Compounds A&B to Compound A (LOQ)

1 mg/mL Cmpd A Spiking Soln x 5.0 mL = 5.0 mg 100% Cmpd A
0.001 mg/mL Related A&B Spiking Soln x 2.5 mL = 0.0025 mg Related A&B
0.0025 mg Related A&B/5.0 mg Cmpd A x 100 = 0.050 w/w %

1-0.10% / 2-0.10% / 3-0.10%

Flasks (each): 150 mg Placebo + 5.0 mL of Cmpd A Spiking Solution + 5.0 mL of Compound A & B Spiking Solution & QS with Diluent — 50 mL Volumetric Flask

0.10 w/w % Related Compounds A&B to Compound A

1 mg/mL Cmpd A Spiking Soln x 5.0 mL = 5.0 mg 100% Cmpd A
0.001 mg/mL Related A&B Spiking Soln x 5.0 mL = 0.0050 mg Related A&B
0.0050 mg Related A&B/5.0 mg Cmpd A x 100 = 0.10 w/w %

1-0.25% / 2-0.25% / 3-0.25%

Flasks (each): 150 mg Placebo + 5.0 mL of Cmpd A Spiking Solution + 12.5 mL of Compound A & B Spiking Solution & QS with Diluent — 50 mL Volumetric Flask

0.25 w/w % Related Compounds A&B to Compound A

1 mg/mL Cmpd A Spiking Soln x 5.0 mL = 5.0mg 100%Cmpd A
0.001 mg/mL Related A&B Spiking Soln x 12.5mL=0.0125 mg Related A&B
0.0125 mg Related A&B/5.0 mg Cmpd A x 100 = 0.25 w/w %

5.0 mg Tablet Weight = 155 mg

150 mg Placebo
+
5.0 mg Compound A

= 155 mg Tablet

Compound A Spiking Solution

100 mg Compound A x 1/100 mL = 1.00 mg/mL

Related A and Related B Stock Solution

10 mg Related A + 10 mg B x 1/10 mL = 1 mg/mL A & B

Related A & B Spiking Solution

1 mg/mL Related A & B Stock x 1mL x 1/1000 mL = 0.001 mg/mL A & B

Related compounds A and B accuracy (recovery) sample preparation table page 1 of 2.

ATTACHMENT B—SAMPLE PREPARATION DIAGRAMS

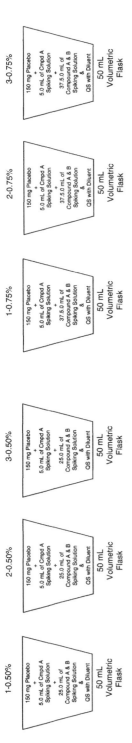

1-0.50%
150 mg Placebo
+
5.0 mL of Cmpd A Spiking Solution
+
25.0 mL of Compound A & B Spiking Solution
&
QS with Diluent
50 mL Volumetric Flask

2-0.50%
150 mg Placebo
+
5.0 mL of Cmpd A Spiking Solution
+
25.0 mL of Compound A & B Spiking Solution
&
QS with Diluent
50 mL Volumetric Flask

3-0.50%
150 mg Placebo
+
5.0 mL of Cmpd A Spiking Solution
+
25.0 mL of Compound A & B Spiking Solution
&
QS with Diluent
50 mL Volumetric Flask

1-0.75%
150 mg Placebo
+
5.0 mL of Cmpd A Spiking Solution
+
37.5.0 mL of Compound A & B Spiking Solution
&
QS with Diluent
50 mL Volumetric Flask

2-0.75%
150 mg Placebo
+
5.0 mL of Cmpd A Spiking Solution
+
37.5.0 mL of Compound A & B Spiking Solution
&
QS with Diluent
50 mL Volumetric Flask

3-0.75%
150 mg Placebo
+
5.0 mL of Cmpd A Spiking Solution
+
37.5.0 mL of Compound A & B Spiking Solution
&
QS with Diluent
50 mL Volumetric Flask

0.50 w/w % Related Compounds A&B to Compound A (Limit)

1mg/mL Cmpd A Spiking Soln x 5.0 mL = 5.0 mg 100% Cmpd A
0.001 mg/mL Related A&B Spiking Soln x 25 mL = 0.025 mg Related A&B
0.025 mg Related A&B/5.0 mg Cmpd A x 100 = 0.50 w/w %

0.75 w/w % Related Compounds A&B to Compound A

1 mg/mL Cmpd ASpiking Soln x5.0 mL = 5.0mg 100%CmpdA
0.001 mg/mL Related A&B Spiking Soln x 37.5 mL = 0.0375 mg Related A&B
0.0375 mg Related A&B/5.0 mg Cmpd A x 100 = 0.75 w/w %

Limit for Related A and B = 0.5 w/w% of Compound A

LOQ =0.05%

25% of Limit = 0.10%

50% of Limit = 0.25%

100% of Limit = 0.50%

150% of Limit = 0.75%

5.0 mg Tablet Weight = 155 mg

150 mg Placebo
+
5.0 mg Compound A

= 155 mg Tablet

Compound A Spiking Solution

100 mg Compound A x 1/100mL = 1.00 mg/mL

Related A and Related B Stock Solution

10 mg Related A + 10 mg B x 1/10mL = 1 mg/mL A & B

Related A & B Spiking Solution

1 mg/mL Related A & B Stock x 1 mL x1/1000 mL = 0.001 mg/mL A & B

Related compounds A and B accuracy (recovery) sample preparation table page 2 of 2.

ATTACHMENT B—SAMPLE PREPARATION DIAGRAMS

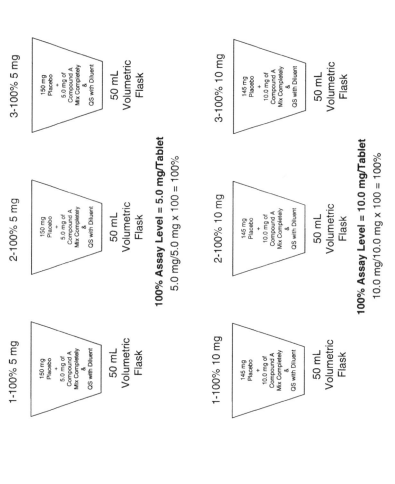

1-100% 5 mg

150 mg
Placebo
+
5.0 mg of
Compound A
Mix Completely
&
QS with Diluent

50 mL
Volumetric
Flask

2-100% 5 mg

150 mg
Placebo
+
5.0 mg of
Compound A
Mix Completely
&
QS with Diluent

50 mL
Volumetric
Flask

3-100% 5 mg

150 mg
Placebo
+
5.0 mg of
Compound A
Mix Completely
&
QS with Diluent

50 mL
Volumetric
Flask

100% Assay Level = 5.0 mg/Tablet
5.0 mg/5.0 mg x 100 = 100%

1-100% 10 mg

145 mg
Placebo
+
10.0 mg of
Compound A
Mix Completely
&
QS with Diluent

50 mL
Volumetric
Flask

2-100% 10 mg

145 mg
Placebo
+
10.0 mg of
Compound A
Mix Completely
&
QS with Diluent

50 mL
Volumetric
Flask

3-100% 10 mg

145 mg
Placebo
+
10.0 mg of
Compound A
Mix Completely
&
QS with Diluent

50 mL
Volumetric
Flask

100% Assay Level = 10.0 mg/Tablet
10.0 mg/10.0 mg x 100 = 100%

5-mg and 10-mg dry spiked placebo sample preparation.

ATTACHMENT B—SAMPLE PREPARATION DIAGRAMS

1-100%

155 mg Grind
&
QS With
Diluent

50 mL
Volumetric
Flask

2-100%

155 mg Grind
&
QS With
Diluent

50 mL
Volumetric
Flask

3-100%

155 mg Grind
&
QS With
Diluent

50 mL
Volumetric
Flask

4-100%

155 mg Grind
&
QS With
Diluent

50 mL
Volumetric
Flask

5-100%

155 mg Grind
&
QS With
Diluent

50 mL
Volumetric
Flask

6-100%

155 mg Grind
&
QS With
Diluent

50 mL
Volumetric
Flask

5 mg Tablet Repeatability and Intermediate Precision Compound A

155 mg Grind x 1 Tablet/155 mg Grind x 5 mg Compound A/Tablet = 5 mg Compound A
5 mg Compound A x 1/50 mL = 0.100 mg/mL Compound A

1-100%

77.5 mg Grind
&
QS With
Diluent

50 mL
Volumetric
Flask

2-100%

77.5 mg Grind
&
QS With
Diluent

50 mL
Volumetric
Flask

3-100%

77.5 mg Grind
&
QS With
Diluent

50 mL
Volumetric
Flask

4-100%

77.5 mg Grind
&
QS With
Diluent

50 mL
Volumetric
Flask

5-100%

77.5 mg Grind
&
QS With
Diluent

50 mL
Volumetric
Flask

6-100%

77.5 mg Grind
&
QS With
Diluent

50 mL
Volumetric
Flask

10 mg Tablet Repeatability and Intermediate Precision Compound A

77.5 mg Grind x 1 Tablet/155 mg Grind x 10 mg Compound A/Tablet = 5 mg Compound A
5 mg Compound A x 1/50 mL = 0.100 mg/mL Compound A

5-mg and 10-mg tablet repeatability and intermediate precision sample preparation for compound A.

ATTACHMENT B—SAMPLE PREPARATION DIAGRAMS

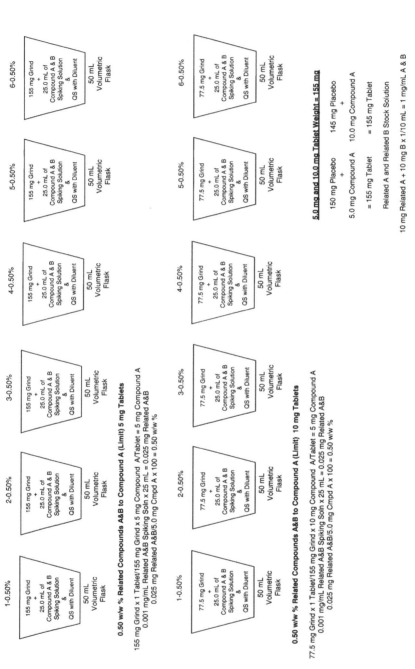

1-0.50%
155 mg Grind
+
25.0 mL of
Compound A & B
Spiking Solution
&
QS with Diluent
→ 50 mL Volumetric Flask

2-0.50%
155 mg Grind
+
25.0 mL of
Compound A & B
Spiking Solution
&
QS with Diluent
→ 50 mL Volumetric Flask

3-0.50%
155 mg Grind
+
25.0 mL of
Compound A & B
Spiking Solution
&
QS with Diluent
→ 50 mL Volumetric Flask

4-0.50%
155 mg Grind
+
25.0 mL of
Compound A & B
Spiking Solution
&
QS with Diluent
→ 50 mL Volumetric Flask

5-0.50%
155 mg Grind
+
25.0 mL of
Compound A & B
Spiking Solution
&
QS with Diluent
→ 50 mL Volumetric Flask

6-0.50%
155 mg Grind
+
25.0 mL of
Compound A & B
Spiking Solution
&
QS with Diluent
→ 50 mL Volumetric Flask

0.50 w/w % Related Compounds A&B to Compound A (Limit) 5 mg Tablets

155 mg Grind x 1 Tablet/155 mg Grind x 5 mg Compound A/Tablet = 5 mg Compound A
0.001 mg/mL Related A&B Spiking Soln x 25 mL = 0.025 mg Related A&B
0.025 mg Related A&B/5.0 mg Cmpd A x 100 = 0.50 w/w %

1-0.50%
77.5 mg Grind
+
25.0 mL of
Compound A & B
Spiking Solution
&
QS with Diluent
→ 50 mL Volumetric Flask

2-0.50%
77.5 mg Grind
+
25.0 mL of
Compound A & B
Spiking Solution
&
QS with Diluent
→ 50 mL Volumetric Flask

3-0.50%
77.5 mg Grind
+
25.0 mL of
Compound A & B
Spiking Solution
&
QS with Diluent
→ 50 mL Volumetric Flask

4-0.50%
77.5 mg Grind
+
25.0 mL of
Compound A & B
Spiking Solution
&
QS with Diluent
→ 50 mL Volumetric Flask

5-0.50%
77.5 mg Grind
+
25.0 mL of
Compound A & B
Spiking Solution
&
QS with Diluent
→ 50 mL Volumetric Flask

6-0.50%
77.5 mg Grind
+
25.0 mL of
Compound A & B
Spiking Solution
&
QS with Diluent
→ 50 mL Volumetric Flask

0.50 w/w % Related Compounds A&B to Compound A (Limit) 10 mg Tablets

77.5 mg Grind x 1 Tablet/155 mg Grind x 10 mg Compound A/Tablet = 5 mg Compound A
0.001 mg/mL Related A&B Spiking Soln x 25 mL = 0.025 mg Related A&B
0.025 mg Related A&B/5.0 mg Cmpd A x 100 = 0.50 w/w %

5.0 mg and 10.0 mg Tablet Weight = 155 mg

150 mg Placebo   145 mg Placebo
+   +
5.0 mg Compound A   10.0 mg Compound A

= 155 mg Tablet   = 155 mg Tablet

Related A and Related B Stock Solution

10 mg Related A + 10 mg B x 1/10 mL = 1 mg/mL A & B

Related A & B Spiking Solution

1 mg/mL Related A & B Stock x 1mL x 1/1000 mL = 0.001 mg/mL· A & B

5-mg and 10-mg tablet repeatability and intermediate precision sample preparation for related A and B.

ATTACHMENT B—SAMPLE PREPARATION DIAGRAMS

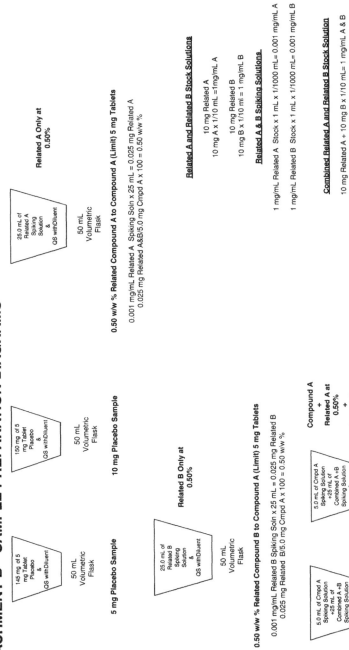

145 mg of 5 mg Tablet Placebo & QS with Diluent

50 mL Volumetric Flask

5 mg Placebo Sample

150 mg of 5 mg Tablet Placebo & QS with Diluent

50 mL Volumetric Flask

10 mg Placebo Sample

25.0 mL of Related B Spiking Solution & QS with Diluent

50 mL Volumetric Flask

Related B Only at 0.50%

25.0 mL of Related A Spiking Solution & QS with Diluent

50 mL Volumetric Flask

Related A Only at 0.50%

0.50 w/w % Related Compound B to Compound A (Limit) 5 mg Tablets

0.001 mg/mL Related B Spiking Soln x 25 mL = 0.025 mg Related B
0.025 mg Related B/5.0 mg Cmpd A x 100 = 0.50 w/w %

0.50 w/w % Related Compound A to Compound A (Limit) 5 mg Tablets

0.001 mg/mL Related A Spiking Soln x 25 mL = 0.025 mg Related A
0.025 mg Related A&B/5.0 mg Cmpd A x 100 = 0.50 w/w %

5.0 mL of Cmpd A Spiking Solution +25 mL of Combined A +B Spiking Solution & QS with Diluent

50 mL Volumetric Flask

5.0 mL of Cmpd A Spiking Solution +25 mL of Combined A +B Spiking Solution & QS with Diluent

50 mL Volumetric Flask

Compound A + Related A at 0.50% + Related B at 0.50%

100% Assay Level = 5.0 mg/Tablet
1mg/mL x 5.0 mL = 5.0 mg/5.0 mg x 100 = 100%
+
Related A and B at the 0.50% Limit
0.001mg/mL x 25.0 mL =0.025mg/5.0 mg x 100 = 0.50% w/w

Related A and Related B Stock Solutions

10 mg Related A
10 mg A x 1/10 mL =1mg/mL A

10 mg Related B
10 mg B x 1/10 ml = 1 mg/mL B

Related A & B Spiking Solutions

1 mg/mL Related A Stock x 1 mL x 1/1000 mL= 0.001 mg/mL A
1 mg/mL Related B Stock x 1 mL x 1/1000 mL= 0.001 mg/mL B

Combined Related A and Related B Stock Solution

10 mg Related A + 10 mg B x 1/10 mL= 1 mg/mL A & B

Combined Related A & B Spiking Solution

1 mg/mL Related A & B Stock x 1 mL x1/1000 mL = 0.001 mg/mL A &B

Compound A Spiking Solution

100 mg Compound A x 1/100 mL = 1.00 mg/mL

Selectivity and specificity sample preparation.

ATTACHMENT B—SAMPLE PREPARATION DIAGRAMS

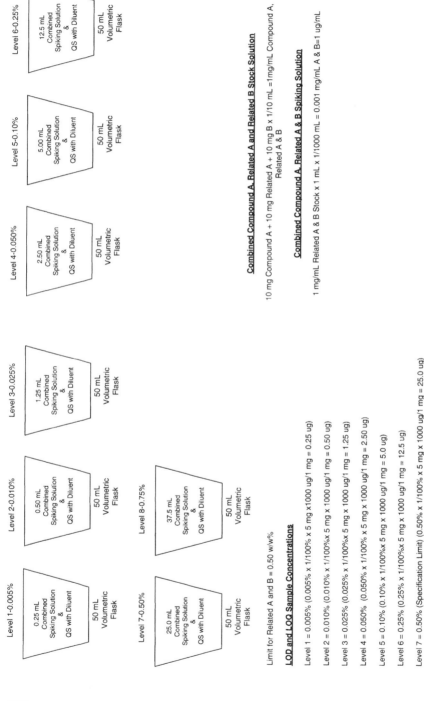

Level 1-0.005% — 0.25 mL Combined Spiking Solution & QS with Diluent — 50 mL Volumetric Flask

Level 2-0.010% — 0.50 mL Combined Spiking Solution & QS with Diluent — 50 mL Volumetric Flask

Level 3-0.025% — 1.25 mL Combined Spiking Solution & QS with Diluent — 50 mL Volumetric Flask

Level 4-0.050% — 2.50 mL Combined Spiking Solution & QS with Diluent — 50 mL Volumetric Flask

Level 5-0.10% — 5.00 mL Combined Spiking Solution & QS with Diluent — 50 mL Volumetric Flask

Level 6-0.25% — 12.5 mL Combined Spiking Solution & QS with Diluent — 50 mL Volumetric Flask

Level 7-0.50% — 25.0 mL Combined Spiking Solution & QS with Diluent — 50 mL Volumetric Flask

Level 8-0.75% — 37.5 mL Combined Spiking Solution & QS with Diluent — 50 mL Volumetric Flask

Combined Compound A, Related A and Related B Stock Solution

10 mg Compound A + 10 mg Related A + 10 mg B x 1/10 mL =1mg/mL Compound A, Related A & B

Combined Compound A, Related A & B Spiking Solution

1 mg/mL Related A & B Stock x 1 mL x 1/1000 mL = 0.001 mg/mL A & B=1 ug/mL

Limit for Related A and B = 0.50 w/w%

LOD and LOQ Sample Concentrations

Level 1 = 0.005% (0.005% x 1/100% x 5 mg x1000 ug/1 mg = 0.25 ug)

Level 2 = 0.010% (0.010% x 1/100%x 5 mg x 1000 ug/1 mg = 0.50 ug)

Level 3 = 0.025% (0.025% x 1/100%x 5 mg x 1000 ug/1 mg = 1.25 ug)

Level 4 = 0.050% (0.050% x 1/100%x 5 mg x 1000 ug/1 mg = 2.50 ug)

Level 5 = 0.10% (0.10% x 1/100%x 5 mg x 1000 ug/1 mg = 5.0 ug)

Level 6 = 0.25% (0.25% x 1/100%x 5 mg x 1000 ug/1 mg = 12.5 ug)

Level 7 = 0.50% (Specification Limit) (0.50% x 1/100% x 5 mg x 1000 ug/1 mg = 25.0 ug)

Level 8 = 0.75% (150% of Specification Limit) (0.75% x 1/100% x 5 mg x 1000 ug/1 mg = 37.5 ug)

Related compounds A and B limit of detection and limit of quantitation sample preparation.

ATTACHMENT B—SAMPLE PREPARATION DIAGRAMS

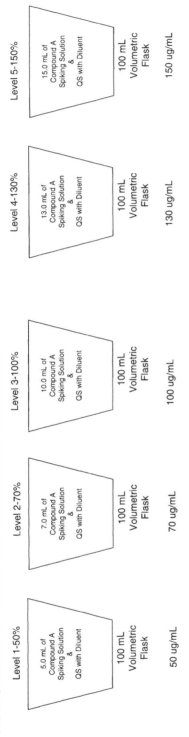

Level 1-50%

5.0 mL of
Compound A
Spiking Solution
&
QS with Diluent

100 mL
Volumetric
Flask

50 ug/mL

Level 2-70%

7.0 mL of
Compound A
Spiking Solution
&
QS with Diluent

100 mL
Volumetric
Flask

70 ug/mL

Level 3-100%

10.0 mL of
Compound A
Spiking Solution
&
QS with Diluent

100 mL
Volumetric
Flask

100 ug/mL

Level 4-130%

13.0 mL of
Compound A
Spiking Solution
&
QS with Diluent

100 mL
Volumetric
Flask

130 ug/mL

Level 5-150%

15.0 mL of
Compound A
Spiking Solution
&
QS with Diluent

100 mL
Volumetric
Flask

150 ug/mL

Compound A Spiking Solution

100 mg Compound A x 1/100 mL = 1.00 mg/mL

Nominal Standard Concentration for Assay = 10.0 mg x 1/100 mL x 1000 ug/1mg =100 ug/mL ofCompound A

Linearity Sample Concentrations

Level 1 = 50% (5 ml x1.00 mg/mL x 1/100ml x 1000 ug/mg = 50 ug/mL X1/100 ug/mL x 100 = 50%)

Level 2 = 70% (7 ml x 1.00 mg/mL x 1/100 mlx 1000 ug/mg = 70 ug/mL X 1/100 ug/mL x 100 = 70%)

Level 3 = 100% (10 ml x 1.00 mg/mL x 1/100 ml x1000 ug/mg = 100 ug/mL X1/100 ug/mL x 100 = 100%)

Level 4 = 130% (13 ml x 1.00 mg/mL x 1/100 ml x1000 ug/mg = 130 ug/mL X 1/100 ug/mL x 100 = 130%)

Level 5 = 150% (15 ml x 1.00 mg/mL x 1/100 ml x1000 ug/mg = 150 ug/mL X1/100 ug/mL x 100 = 150%)

Linearity sample preparation (assay).

211

ATTACHMENT B—SAMPLE PREPARATION DIAGRAMS

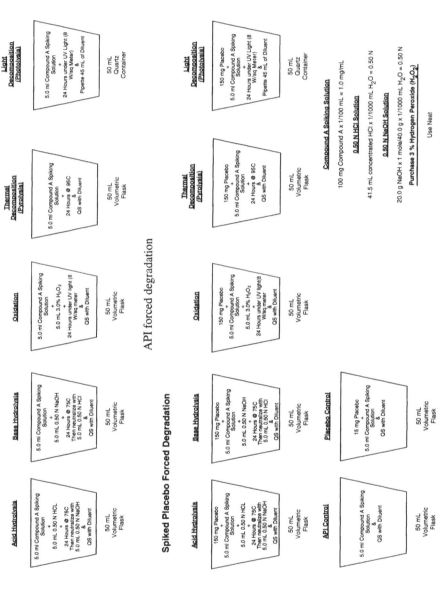

API forced degradation

Acid Hydrolysis

5.0 ml Compound A Spiking Solution
&
5.0 mL 0.50 N HCL

24 Hours @ 75C
Then neutralize with
5.0 mL 0.50 N NaOH
&
QS with Diluent

50 mL
Volumetric
Flask

Base Hydrolysis

5.0 ml Compound A Spiking Solution
+
5.0 mL 0.50 N NaOH

24 Hours @ 75C
Then neutralize with
5.0 mL 0.50 N HCl
&
QS with Diluent

50 mL
Volumetric
Flask

Oxidation

5.0 ml Compound A Spiking Solution
+
5.0 mL 3.0% H₂O₂
+
24 Hours under UV light (8 W/sq meter)

QS with Diluent

50 mL
Volumetric
Flask

Thermal Decomposition (Pyrolysis)

5.0 ml Compound A Spiking Solution

24 Hours @ 95C
&
QS with Diluent

50 mL
Volumetric
Flask

Light Decomposition (Photolysis)

5.0 ml Compound A Spiking Solution
+
24 Hours under UV Light (8 W/sq Meter)
&
Pipette 45 mL of Diluent

50 mL
Quartz
Container

Spiked Placebo Forced Degradation

Acid Hydrolysis

150 mg Placebo
+
5.0 ml Compound A Spiking Solution
+
5.0 mL 0.50 N HCL

24 Hours @ 75C
Then neutralize with
5.0 mL 0.50 N NaOH
&
QS with Diluent

50 mL
Volumetric
Flask

Base Hydrolysis

150 mg Placebo
+
5.0 ml Compound A Spiking Solution
+
5.0 mL 0.50 N NaOH

24 Hours @ 75C
Then neutralize with
5.0 mL 0.50 N HCl
&
QS with Diluent

50 mL
Volumetric
Flask

Oxidation

150 mg Placebo
+
5.0 ml Compound A Spiking Solution
+
5.0 mL 3.0% H₂O₂
+
24 Hours under UV light (8 W/sq meter)
&
QS with Diluent

50 mL
Volumetric
Flask

Thermal Decomposition (Pyrolysis)

150 mg Placebo
+
5.0 ml Compound A Spiking Solution
+
24 Hours @ 95C
&
QS with Diluent

50 mL
Volumetric
Flask

Light Decomposition (Photolysis)

150 mg Placebo
+
5.0 ml Compound A Spiking Solution
+
24 Hours under UV Light (8 W/sq Meter)
&
Pipette 45 mL of Diluent

50 mL
Quartz
Container

API Control

5.0 ml Compound A Spiking Solution
&
QS with Diluent

50 mL
Volumetric
Flask

Placebo Control

15 mg Placebo
+
5.0 ml Compound A Spiking Solution
&
QS with Diluent

50 mL
Volumetric
Flask

Compound A Spiking Solution

100 mg Compound A x 1/100 mL = 1.0 mg/mL

0.50 N HCl Solution

41.5 mL concentrated HCl x 1/1000 mL H₂O = 0.50 N

0.50 N NaOH Solution

20.0 g NaOH x 1 mole/40.0 g x 1/1000 mL H₂O = 0.50 N

Purchase 3 % Hydrogen Peroxide (H₂O₂)

Use Neat

Spiked placebo forced degradation

212

ATTACHMENT B—SAMPLE PREPARATION DIAGRAMS

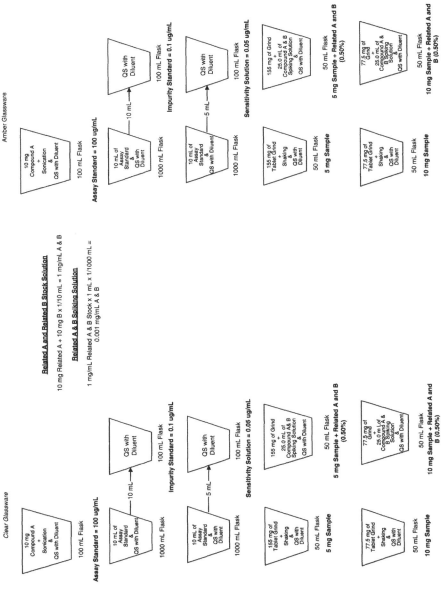

Ambient temperature solution stability.

ATTACHMENT B—SAMPLE PREPARATION DIAGRAMS

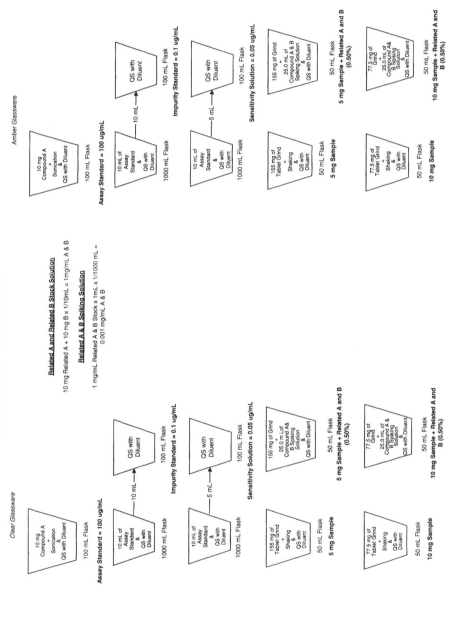

Refrigerated temperature solution stability.

ATTACHMENT B—SAMPLE PREPARATION DIAGRAMS

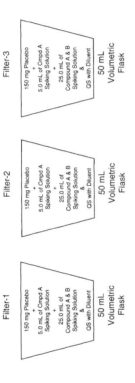

Filter-1

150 mg Placebo
+
5.0 mL of Cmpd A Spiking Solution
+
25.0 mL of Compound A & B Spiking Solution
&
QS with Diluent

50 mL Volumetric Flask

Filter-2

150 mg Placebo
+
5.0 mL of Cmpd A Spiking Solution
+
25.0 mL of Compound A & B Spiking Solution
&
QS with Diluent

50 mL Volumetric Flask

Filter-3

150 mg Placebo
+
5.0 mL of Cmpd A Spiking Solution
+
25.0 mL of Compound A & B Spiking Solution
&
QS with Diluent

50 mL Volumetric Flask

0.50 w/w % Related Compounds A&B to Compound A (Limit)

1 mg/mL Cmpd A Spiking Soln x 5.0 mL = 5.0 mg 100% Cmpd A
0.001 mg/mL Related A&B Spiking Soln x 25 mL = 0.025 mg Related A&B
0.025 mg Related A&B/5.0 mg Cmpd A x 100 = 0.50 w/w %

5.0 mg Tablet Weight = 155 mg

150 mg Placebo
+
5.0 mg Compound A

= 155 mg Tablet

Compound A Spiking Solution

100 mg Compound A x 1/100 mL = 1.00 mg/mL

Related A and Related B Stock Solution

10 mg Related A + 10 mg B x 1/10 mL = 1 mg/mL A & B

Related A & B Spiking Solution

1 mg/mL Related A & B Stock x 1 mL x 1/1000 mL = 0.001 mg/mL A & B

Filter retention study sample preparation.

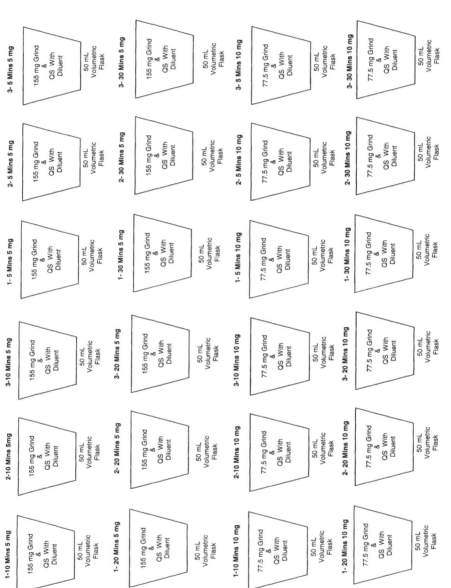

5-mg and 10-mg tablet grind extraction efficiency study sample preparation: sonication time variation.

Notes

APPENDIX VII

TEMPLATE FOR AN EXAMPLE METHODS VALIDATION REPORT

Assay and Impurities Methods Validation for Compound A in Your Product Tablets, 5- and 10-mg

For

Your Company, Inc.

Report No.: R124356

Protocol No.: P124356

Note: All data within this template-example report are dummy values and any correlation to other data, either living or dead, is purely coincidental. Data, calculations, and graphs may not necessarily be reproducible by manual calculation but are included for template format and example purposes only.

Month, Day, Year

Validating Chromatographic Methods. By David M. Bliesner
Copyright © 2006 John Wiley & Sons, Inc.

TABLE OF CONTENTS

Quality Assurance Department Review		
Report Title:	Assay and Impurities Methods Validation for Compound A in Your Product Tablets, 5 and 10 mg	
Report No.:	R124356	
Protocol No.:	P124356	

AUDIT	DATE	STATUS
Raw Data Audit	Month, Day, Year	Complies with GMP
Report Audit and Data Transcription Verification	Month, Day, Year	Complies with GMP

Last Name, First Name, Title   Date

Methods Validation Report Approval	
Report Title:	Assay, Impurities and Content Uniformity Methods Validation for Compound A in Your Product Tablets, 5 and 10 mg
Report No.:	R124356
Protocol No.:	P124356

Author:

_____
First Name, Last Name, Title   Date

Reviewer:

_____
First Name, Last Name, Title   Date

Approval:

_____
First Name, Last Name, Title   Date

I. INTRODUCTION

The active pharmaceutical ingredient (API) contained in your products 5-mg and 10-mg tablets is compound A and possesses the following molecular structure and weight:

$C_{13}H_{17}N$ MW 187.29

Compound A's API is manufactured by the following synthetic process:

Compound A possess two potential impurities: related A and related B. Related A is a process impurity and related B is both a process impurity and potential degradant. Molecular structures and weights of both are given below.
Related A possesses the following molecular structure and weight:

$C_{10}H_{15}N$ MW 149.23

Related B possesses the following molecular structure:

$C_{11}H_{13}N$ MW 159.23

Degradation occurs through exposure to light via the following proposed mechanism:

| Compound A | Related A | Related B |

II. PURPOSE

Your laboratory has validated a combined assay and impurities method for analysis of compound A your product tablets, 5 mg and 10 mg. The validation was conducted according to the method developed at your lab and reported in D124356, "Method Development Report for the Analysis of Compound A and Related Compounds A and B in Drug Product," and "Validation of Analytical Test Procedures," SOP Number ABC-1243 rev4 dated 1 April 2002.

III. EXPERIMENTAL

A. Reagents

Sodium acetate, Tisher, HPLC grade

Glacial acetic acid, Tisher, HPLC grade

Acetonitrile, Tisher, Suptima grade

THF, Tisher, Suptima grade

Water, in-house deionized water

B. Samples

Compound A , micronized, lot 1, purity 100.2%

Compound A , 5-mg tablets, lot 2, aged samples

Compound A , 10-mg tablets, lot 3, aged samples

Compound A , 5-mg tablets, lot 4

Compound A , 10-mg tablets, lot 5

Compound A placebo, (5.0 mg), lot 6

Compound A placebo, (10.0 mg), lot 7

FD & C Red #40 AL-LAKE, lot Z

C. Standards

Compound A reference standard, USP lot K 100.0%

Related A lot 200109240001, purity 98.6%

Related B lot RS990002711, purity 99.25%

D. Instrumentation

HPLC pump: Serta, model 2666 & 2667; Hewey Dewey, model 0001

UV detector: Serta, model 2387; SP model UV100

PDA detector: as part of a Hewey Dewey system, model 0002

Autosampler: Serta, model 2590 & 2595; SP model AS200

IV. RESULTS AND DISCUSSION

All raw data can be found in cited notebook reference locations.

a. Linearity

i. Assay

Assay linearity was demonstrated by preparing five standard solutions within the range of ~50% to 150% of the nominal sample concentration (0.2 mg/mL). Each solution was prepared by serial dilution from a single stock and was injected in triplicate. Linear regression analysis was performed, excluding the origin as a point. Results are presented in Table 1. All acceptance criteria were met. A graph of concentration versus area response is shown in Figure 1. A graph of the residuals versus analyte concentration is shown in Figure 2. No trend was observed from the residual plot. The y intercept did not show significant departure from zero.

TABLE 1   Linearity of Compound A for Assay

Linearity of Compound A for Assay

Target Level	Concentration (μg/mL)	Area Response	Average	% RSD
50%	101.1	2601308.00 2602304.50 2605056.00	2602890	0.075%
70%	141.2	3632631.50 3635443.50 3633128.50	3633728	0.041%
100%	201.0	5174038.25 5172909.50 5175717.25	5174222	0.027%
130%	263.3	6747457.50 6764901.00 6752018.00	6754972	0.143%
150%	302.5	7751256.00 7757338.50 7763704.25	7757433	0.080%
Slope:			25640.9	
Intercept:			14889.2	
Correlation coefficient (r^2):			0.99997	

Reference: LAB BOOK 15 OWNER pp. 22, 18, Figures 1 and 2.
Specification: The correlation coefficient (r^2) should be $\geq$ 0.999.

ii. Impurities

Impurities linearity was demonstrated for each impurity and for parent compound A to lower levels (data of which were used to calculate relative response factors and validity of relative impurity standard range) by preparing five standard solutions within the range of ~0.05% (~LOQ) to 150% of the impurity specification (0.5% or 1 µg/mL). Each solution was prepared by serial dilution from a single stock and was injected five times. Linear regression analysis was performed, excluding the origin as a point. Results are presented in Tables 2–4. Plots of concentration versus area response and residual plots versus analyte concentration are shown in Figures 3–8. All acceptance criteria were met. No trends were observed from the residual plots. The *y* intercept did not show significant departure from zero.

b. Range

i. Assay

Range for the assay method was demonstrated by analyzing placebo solutions spiked in a range between ~50% and 150% of the nominal method sample concentration of compound A (0.2 mg/mL) Three weights were prepared at each of five concentration levels and each solution was analyzed in triplicate. Linear regression analysis was performed, excluding the origin as a point. Results are presented in Table 5. All acceptance criteria were met. A graph of concentration versus area response is shown in Figure 9. A graph of the residuals versus analyte concentration is shown in Figure 10. No trend was observed from the residual plot. The *y* intercept did not show significant departure from zero.

ii. Impurities

Range for the impurities method was demonstrated by analyzing placebo solutions spiked in a range between approximately LOQ (~0.05% of the nominal assay concentration, or 0.1 µg/mL) to 150% of the impurity-specification. Three weights were prepared at each of five concentration levels and each solution was analyzed in triplicate. Linear regression analysis was performed, excluding the origin as a point. The *y* intercept did not show significant departure from zero. Tables 6–7 present the results for impurities related A and related B, respectively. Plots of concentration versus area response and residual plots versus analyte concentration are shown in Figures 11–14. The acceptance criteria were met for all impurities.

TABLE 2   Linearity of Parent Compound A at Low Levels

		Linearity of Parent Compound A At Low Levels		
Target Level	Concentration (mg/mL)	Area Response	Average	% RSD
0.05%	0.1083	3139.00 2597.50 3073.00 3129.75 3220.50	3031.95	8.20%
0.125%	0.2067	6645.50 5619.00 6477.50 6190.00 6155.50	6317.50	6.30%
0.25%	0.5109	14057.50 14293.75 14452.75 14357.75 14576.50	14437.65	1.35%
0.50%	1.0380	29258.50 28897.50 29271.50 28949.00 29293.00	29133.90	0.66%
0.75%	1.4828	41814.00 41709.00 41489.50 41324.50 41156.00	41500.40	0.65%
Slope:				27856.28
Intercept:				200.95
Correlation coefficient (r^2):				0.99899

Reference: LAB BOOK 22 OWNER pp. 14, 15, 22A, 18A, 20, 44, 46.
Linearity and residual plots: Figures 3 and 4.
Specification: The correlation coefficient (r^2) should be ≥ 0.995.

Overall, the method was determined to have a range of 50% to 150% of the nominal concentration. The range for impurities was established from 0.05% to 0.75% of the nominal concentration for both related A and B

TABLE 3   Linearity of Related A

		Linearity of Related A		
Target Level	Concentration (mg/mL)	Area Response	Average	% RSD
0.05%	0.1064	2722.50 2715.00 2601.00 2689.50 2619.50	2669.50	2.09%
0.10%	0.2128	5055.00 5187.75 5015.50 5197.00 5138.00	5118.65	1.57%
0.25%	0.5321	13003.00 12895.00 12900.50 12771.50 12825.00	12879.00	0.68%
0.50%	1.0642	26301.75 26355.75 26113.00 26307.50 26101.50	26175.90	0.25%
0.75%	1.5203	37313.50 36866.00 37045.50 36930.50 37061.25	37043.35	0.46%
Slope:				24431.05
Intercept:				−10.76
Correlation coefficient (r^2):				0.99992

Reference: LAB BOOK 22 OWNER, pp. 14, 15A, 22, 18, 20, 47, 49.
Linearity and residual plots: Figures 5 and 6.
Specification: The correlation coefficient (r^2) should be ≥ 0.995.

The specifications: The minimum assay range should be at least 80%–120% of the sample concentration. The minimum content uniformity range should be at least 70%–130% of the sample concentration. The impurities/degradants range should be from the LOQ–150% of the impurities/degradants specification, were met.

TABLE 4   Linearity of Related B

Linearity of Related B				
Target Level	Concentration (mg/mL)	Area Response	Average	% RSD
0.05%	0.1037	3230.50 3254.50 3377.75 3143.00 3297.00	3258.75	2.74%
0.10%	0.2075	6360.50 6447.25 6455.25 6328.50 6403.25	6398.95	0.85%
0.25%	0.5187	22039.00 22025.50 22144.00 15890.00 22208.50	20861.40	13.33%
0.50%	1.0374	32557.50 32573.75 32230.75 32386.00 32312.00	32412.00	0.46%
0.75%	1.4820	45760.50 46056.75 45797.50 45763.25 45839.00	45843.40	0.27%
Slope:				30467.78
Intercept:				1435.75
Correlation coefficient (r^2):				0.9961

Reference: LAB BOOK 22 OWNER, pp. 14, 15A, 22, 18, 20, 50, 52.
Linearity and residual plots: Figures 7 and 8.
Specification: The correlation coefficient (r^2) should be ≥ 0.995.

c. Accuracy (Recovery)

i. Assay

Accuracy and recovery of the method for assay was demonstrated by analyzing data obtained from spiked placebo solutions during the range portion of

TABLE 5   Range of the Method Compound A

Target Level	Concentration (mg/mL)	Average Areas ($n = 3$)	% RSD
50%	100.70	2710261.50	0.50
	99.70	2690415.67	0.01
	99.80	2684382.67	0.02
70%	140.89	3785161.00	0.03
	139.58	3751891.67	0.02
	139.72	3746069.50	0.09
100%	201.40	5407192.33	0.07
	199.40	5342433.67	0.20
	199.60	5430952.17	0.03
130%	261.82	7042352.83	0.07
	257.22	6943523.00	0.06
	257.48	6943275.67	0.11
150%	298.10	8064244.67	0.05
	299.10	8004663.67	0.02
	299.40	8087111.17	0.08
Slope:			26690335.5
Intercept:			20103.931
Correlation coefficient (r^2):			0.9999

Reference: LAB BOOK 15 OWNER pp. 25A and 29A. Figures 9 and 10.
Specification: The Correlation Coefficient (r^2) should be ≥ 0.997.

the validation. The percent recovery of each individual sample weight and the average at each concentration level was determined and is presented in Table 8. Accuracy was performed on the 5-mg tablets only, since the 10-mg strength is a scale up of the 5-mg strength and the 5-mg strength represents the worst case scenario with respect to excipients and potential recovery interferences. All acceptance criteria were met. After examination of the y intercept, residual plot, and recovery data, the method was determined to have no bias.

ii. Impurities

The accuracy and recovery of the method for impurities was demonstrated by analyzing spiked placebo solutions during the range portion of the validation. The percent recovery of each individual impurity weight and the average at each concentration level was determined. Tables 9 and 10 present the recovery data for related A and B, respectively.

TABLE 6   **Range of the Method Related A**

	Range of the Method Related A		
Target Level	Concentration (mg/mL)	Average Area ($n = 3$)	% RSD
	113.6	2696.83	6.08
0.05%	108.9	2489.00	4.61
	104.2	2417.25	2.06
	227.3	5674.75	0.96
0.10%	217.9	5343.92	1.06
	208.4	5225.50	1.01
	568.2	14437.50	0.52
0.25%	544.7	14330.33	0.26
	521.0	12975.67	1.20
	136.3	29375.38	0.99
0.50%	1089.4	27182.25	0.29
	1042.0	26434.77	1.70
	1633.3	40961.53	0.15
0.75%	1556.3	39066.86	0.88
	1488.5	37519.57	0.41
Slope:			25.4655578
Intercept:			22.91999
Correlation coefficient (r^2):			0.99908

Reference: LAB BOOK 19 OWNER pp. 22 and 18.
Linearity and residual plots: Figures 11 and 12.
Specification: The correlation coefficient (r^2) should be ≥ 0.995.

iii. Recovery from Dry Spiked Placebo

In addition to the solution spiked placebo, recovery from dry spiked placebo was also performed for the assay analysis. Three replicate spiked placebos were used to evaluate the recovery from dry spiked placebos. Each placebo was spiked with dry active at 100% of the nominal concentration. These three preparations were centrifuged, injected three times, and analyzed according to the analytical method. The percent recovery and % RSD were determined for each individual sample. Results are presented in Table 11. The specifications were met.

TABLE 7   Range of the Method Related B

| | Range of the Method Related B | | |
Target Level	Concentration (μg/mL)	Average Area ($n = 3$)	% RSD
	104.2	3031.58	2.24
0.05%	106.7	3185.92	1.58
	107.9	3235.00	3.20
	208.3	6373.33	0.63
0.10%	213.5	6499.00	0.69
	215.8	6976.75	1.31
	520.9	22103.17	0.45
0.25%	533.7	22578.08	0.32
	539.4	22720.67	0.35
	1041.7	32184.58	0.17
0.50%	1067.4	33046.33	0.29
	1078.8	33897.92	0.60
	1488.1	46378.50	0.29
0.75%	1524.9	47311.08	0.26
	1541.1	48051.00	0.21
Slope:			300000.86
Intercept:			728.64
Correlation coefficient(r^2):			0.99677

Reference: LAB BOOK 19 OWNER pp. 19 and 21.
Linearity and residual plots: Figures 13 and 14.
Specification: the correlation coefficient (r^2) should be ≥ 0.995.

d. Filter Retention Study

i. Assay

Three replicate spiked placebos were used to evaluate two prospective filters. The 5-mg strength placebo with dye was used throughout the validation since it represents the worst case and the 5-mg and 10-mg placebo are the same except for the dye. Each placebo was spiked with dry active at 100% of the nominal concentration and with each impurity at its specification limit (0.5%). These three preparations were used for the filter study of both assay and impurity. Portions of each sample were filtered through 0.45-μm PVDF and nylon

TABLE 8   Accuracy Compound A

| | Accuracy Compound A | | |
Level	Concentration (mg/mL)	% Recovery	Average % Recovery
50%	100.70	102.0	
	99.70	102.2	102.0
	99.80	101.9	
70%	140.89	101.7	
	139.58	101.8	101.7
	139.72	101.6	
100%	201.40	101.7	
	199.40	101.3	101.5
	199.60	101.4	
130%	261.82	101.8	
	257.22	101.5	101.5
	257.48	101.2	
150%	302.10	101.1	
	299.10	101.4	101.2
	299.40	101.1	

Reference: LAB BOOK 15 OWNER pp. 28A and 29.
Specification: The percentage recovery of the spiked placebos should be 100 ± 2.0% for the average of each set of three weights. Each individual sample recovery should lie within the range of 98%–102%.

syringe filters using a 5-mL flush volume. The filtered aliquots were compared to unfiltered centrifuged portions of the samples (5000 rpm for about 5 minutes). All samples were injected three times and analyzed according to the analytical method. The percent recovery and % RSD were determined for each individual sample. The average result for each filter was compared to the centrifuged sample. Results are presented in Table 12. Both the nylon and PVDF filters were shown to be acceptable for use with this method.

ii. Impurity

The three samples from the assay filter study [see (i)] were used. Portions of each sample were filtered through 0.45-μm PVDF and nylon syringe filters using a 5-mL flush volume. The filtered aliquots were compared to unfiltered, centrifuged portions of the samples. All samples were injected three times and analyzed according to the analytical method. The percent recovery and % RSD were determined for each individual sample. The average

TABLE 9   Accuracy Related A

| | Accuracy Related A | | |
Level	Concentration (mg/mL)	% Recovery	Average % Recovery
0.05%	113.6	100	
	108.9	96	98
	104.2	97	
0.10%	227.3	105	
	217.9	103	104
	208.4	105	
0.25%	568.2	106	
	544.7	104	104
	521.0	103	
0.50%	1136.3	109	
	1089.4	105	107
	1042.0	106	
0.75%	1633.3	106	
	1556.3	105	106
	1488.5	106	

Reference: LAB BOOK 19 OWNER, p. 22.
Specification: The individual and average percent recovery of each impurity should be within the range of 75%–125% at each concentration level.

result for each filter was compared to the centrifuged sample. Results are presented in Tables 13–14 for impurities related A and B, respectively. Both the nylon and PVDF filters were shown to be acceptable for use with this method.

e. Precision

i. Repeatability

a. Assay

The repeatability of the method for assay was demonstrated by preparing six samples for both tablet strengths. The samples were analyzed according to the analytical method and the percent label claim for compound A was determined for each sample. Results are presented in Table 15. All acceptance criteria were met. The chromatograms of a typical standard and sample solutions (5 mg and 10 mg) as well as those spiked with impurities are shown in Figures 15–23, respectively.

TABLE 10   Accuracy Related B

	Accuracy Related B		
Level	Concentration (mg/mL)	% Recovery	Average % Recovery
0.05%	104.2	97	
	106.7	99	99
	107.9	100	
0.10%	208.3	102	
	213.5	101	103
	215.8	102	
0.25%	520.9	103	
	533.7	103	103
	539.4	103	
0.50%	1041.7	103	
	1067.4	103	103
	1078.8	104	
0.75%	1488.1	104	
	1524.9	103	104
	1541.1	104	

Reference: LAB BOOK 19 OWNER, p. 19.
Specification: The individual and average percent recovery of each impurity should be within the range of 75%–125% at each concentration level.

TABLE 11   Dry Spiked Placebos (Compound A)

	Sample Name	Mean % Recovery ($n = 3$ injections)	% RSD	Average % Recovery ($n = 9$ injections)
		Dry Spiked Placebos (Compound A)		
Centrifuged	Sample 1	99.3	4.4	
	Sample 2	99.9	3.7	100.3
	Sample 3	101.8	0.0	

Reference: LAB BOOK 22 OWNE p. 3.
Specification: The percentage recovery of the spiked placebos should be 100 ± 2.0% for the average of each set of three weights. Each individual sample recovery should lie within the range of 98%–102%

b. Impurities

Repeatability of the method for impurities was demonstrated by preparing six samples at both tablet strengths. The samples were then spiked with impurities at 0.1% and analyzed according to the analytical method. The chromatograms

TABLE 12 Filter Study Compound A

		Mean		% Recovery	
Sample Name		% Recovery (n = 3 injections)	% RSD	per Filter (n = 9 injections)	% Difference (vs. centrifuged)
Nylon	Sample 1	101.0	0.4		
	Sample 2	101.0	0.3	101.0	1.0
	Sample 3	101.8	0.1		
PVDF	Sample 1	101.1	0.2		
	Sample 2	100.3	0.1	100.1	0.1
	Sample 3	101.0	0.0		
Centrifuged	Sample 1	100.1	0.2		
	Sample 2	100.3	0.3	100.0	N/A
	Sample 3	100.1	0.1		

Reference: LAB BOOK 19 OWNER p. 51A.
Specification: The percent difference between the filtered sample and the centrifuged sample should be NMT 1.5%. For recovery from dry spiked placebo, the percentage recovery of the spiked placebos should be 100 ± 2.0% for the average of each set of three weights. Each individual sample recovery should lie within the range of 98%–102%.

TABLE 13 Filter Study Related A

		Mean		% Recovery	
Sample Name		% Recovery (n = 3 injections)	% RSD	per Filter (n = 9 injections)	% Difference (vs. centrifuged)
Nylon	Sample 1	106.4	0.4		
	Sample 2	103.4	0.1	105.3	0.5
	Sample 3	106.2	0.4		
PVDF	Sample 1	106.6	0.6		
	Sample 2	103.4	0.4	105.3	0.5
	Sample 3	106.0	0.6		
Centrifuged	Sample 1	106.0	0.5		
	Sample 2	102.9	0.2	104.8	N/A
	Sample 3	105.5	0.5		

Reference: LAB BOOK 19 OWNER p. 52A.
Specification: The percent difference between the filtered sample and the centrifuged sample should be NMT 5%.

of a typical impurity standard (0.1%) and sample solutions (5 mg and 10 mg) spiked with impurities are shown in Figures 17–23, respectively. Results are presented in Tables 16–17.

For both impurities the % RSD values obtained met the specification criteria.

TABLE 14   Filter Study Related B

	Sample Name	Mean % Recovery ($n=3$ injections)	% RSD	% Recovery per Filter ($n=9$ injections)	% Difference (vs. centrifuged)
			Filter Study Related B		
Nylon	Sample 1	101.3	0.3		
	Sample 2	99.4	0.4	100.6	0.3
	Sample 3	101.1	0.2		
PVDF	Sample 1	101.0	0.2		
	Sample 2	99.4	0.5	100.5	0.2
	Sample 3	101.2	0.7		
Centrifuged	Sample 1	100.5	0.5		
	Sample 2	99.1	0.6	100.3	N/A
	Sample 3	101.5	0.3		

Reference: LAB BOOK 19 OWNER p. 53A.
Specification: The percent difference between the filtered sample and the centrifuged sample should be NMT 5%.

TABLE 15   Repeatability Assay (%LC) Analyst 1

Sample Preparation	% LC 5-mg Sample Lot 2	% LC 10-mg Sample Lot 3
	Repeatability Assay (%LC) Analyst 1	
1	89.5	89.2
2	89.6	89.4
3	89.9	99.1
4	89.9	99.1
5	89.7	97.9
6	89.8	89.3
Mean	89.7	89.5
% RSD	0.2	0.5

Reference: LAB BOOK 14 OWNER p. 35A.
Specification: The % RSD of the assay values should not be greater than 2.0%.

ii. Intermediate Precision

a. Assay

Intermediate precision of the method was demonstrated by repeating the repeatability experiment with a second analyst. This analyst used different solution preparations, reagents, operating conditions, column, and HPLC

TABLE 16   Repeatability 5-mg Sample, Lot 2 Impurities (%w/w of Each Impurity)

Repeatability 5-mg Sample, Lot 2 Impurities (%w/w of Each Impurity)			
Sample Preparation	Related A	Related B	Total Impurities
1	0.10	0.09	0.19
2	0.09	0.08	0.17
3	0.09	0.08	0.17
4	0.09	0.08	0.17
5	0.09	0.09	0.18
6	0.08	0.08	0.16
Average	0.09	0.08	0.17
% RSD	6.4	5.9	5.5

Reference: LAB BOOK 14 OWNER p. 35, LAB BOOK 23 OWNER pp. 87 and 88. Specification: The % RSD of the impurities/degradants results should not be greater than 15%.

TABLE 17   Repeatability 10-mg Sample, Lot 3 Impurities (%w/w of Each Impurity)

Repeatability 10-mg Sample, Lot 3 Impurities (%w/w of Each Impurity)			
Sample Preparation	Related A	Related B	Total Impurities
1	0.10	0.09	0.19
2	0.09	0.08	0.17
3	0.09	0.08	0.17
4	0.09	0.08	0.17
5	0.09	0.08	0.17
6	0.08	0.08	0.16
Average	0.09	0.08	0.17
% RSD	6.4	4.6	5.2

Reference: LAB BOOK 14 OWNER p. 35, LAB BOOK 23 OWNER pp. 87 and 88. Specification: The % RSD of the impurities/degradants results should not be greater than 15%.

systems and tested on different days from the first analyst. Individual and combined results for both analysts are presented in Table 18. All acceptance criteria were met.

TABLE 18 Intermediate Precision (Repeatability Analyst 2) Assay (%LC)

	Intermediate Precision (Repeatability Analyst 2) Assay (%LC)	
Sample Preparation	% LC 5-mg Sample Lot 2	% LC 10-mg Sample Lot 3
1	99.4	98.8
2	99.6	98.5
3	99.2	98.4
4	100.0	98.5
5	99.5	98.7
6	99.2	98.7
Mean	99.5	98.6
%RSD	0.30	0.16
Grand mean (Analyst 1 + 2)	99.1	98.6
% RSD ($n = 12$)	0.5	0.4

Reference: LAB BOOK 21 OWNER p. 11.
Specification: The % RSD of the assay values should not be greater than 2.0%. The % RSD of the combined assay values from both analysts should be NMT 3.0%

TABLE 19 Intermediate Precision (Repeatability Analyst 2) 5-mg Sample, Lot 3 Impurities (%w/w of Each Impurity)

	Intermediate Precision (Repeatability Analyst 2) 5-mg Sample, Lot 3 Impurities (%w/w of Each Impurity)		
Sample Preparation	Related A	Related B	Total Impurities
1	0.10	0.09	0.19
2	0.09	0.09	0.18
3	0.09	0.08	0..17
4	0.09	0.09	0.18
5	0.09	0.09	0.18
6	0.09	0.08	0.17
Average	0.09	0.09	0.18
% RSD	0.38	0.20	0.35
Grand average (analyst 1 + 2)			0.18
% RSD ($n = 12$)			0.55

Reference: LAB BOOK 21 OWNER p. 30A.
Specification: The % RSD of the impurities/degradants results should not be greater than 15%. The % RSD of the total percent impurities/degradants over both days is NMT 15%.

TABLE 20   Intermediate Precision (Repeatability Analyst 2) 5-mg Sample, Lot 74401B03 Impurities (%w/w of Each Impurity)

Intermediate Precision (Repeatability Analyst 2) 5-mg Sample, Lot 74401B03 Impurities (%w/w of Each Impurity)

Sample Preparation	Related A	Related B	Total Impurities
1	0.10	0.09	0.19
2	0.09	0.09	0.18
3	0.09	0.08	0..17
4	0.09	0.09	0.18
5	0.09	0.09	0.18
6	0.09	0.08	0.17
Average	0.09	0.09	0.18
% RSD	0.98	0.88	0.51
Grand average (analyst 1 + 2)			0.19
% RSD ($n = 12$)			0.55

Reference: LAB BOOK 21 OWNER p. 30A.
Specification: The % RSD of the impurities/degradants results should not be greater than 15%. The % RSD of the total percent impurities/degradants over both days is NMT 15%.

TABLE 21   Relative Response Factors (RRF = Slope of Compound A/Slope of Impurity.)

Relative Response Factors (RRF = Slope of Compound A/Slope of Impurity.)

Impurity Name	RRF
Related A	1.05
Related B	0.86

Reference: LAB BOOK 22 OWNER p. 43; LAB BOOK 23 OWNER p. 56; LAB BOOK 26 OWNER p. 33.

b. Impurities

Intermediate precision of the method for impurities was demonstrated by duplicating the repeatability experiment with a second analyst. This analyst used different solution preparations, reagents, operating conditions, and tested on different days from the first analyst. Individual and combined results for both analysts are presented in Tables 19–20.

For both impurities the % RSD values obtained met the specification criteria.

TABLE 22   Forced Degradation (Percent Degradation and Peak Purity on the Compound A Peak)

Treatment	Degradation (%)	Peak Purity (Purity Threshold = 990.000)
Control (API and spiked placebo)	N/A	999.866, 999.887
API Acid degradation with heat (24 hrs)	4.4	999.892
API Base degradation with heat (24 hrs)	3.3	999.861
API Peroxide with UV irradiation (24 hrs)	6.2	999.907
API Pyrolysis in aqueous solution (22 hrs)	3.7	999.885
API UV irradiation (48 hrs)	< 2.0	999.886
Spiked placebo acid degradation with heat (24 hrs)	8.4	999.887
Spiked placebo base degradation with heat (24 hrs)	12.6	999.909
Spiked placebo peroxide with UV irradiation (24 hrs)	< 2.0	999.904
Spiked placebo pyrolysis in aqueous solution (22 hrs)	32.2	999.856
Spiked placebo UV irradiation (48 hrs)	< 2.0	999.884

Reference: LAB BOOK 11 OWNER pp. 23, 43–54.
Specification: The purity value is within the allocated peak threshold.

f. Selectivity (Specificity)

i. Matrix (Placebo)/Diluent Interference

1. Two 1X and two 2X placebo solutions were prepared in the presence of dye for the 5-mg dosage strength (worse case). Each placebo preparation was injected in duplicate. Example chromatograms of diluent blank, 1X placebo, and 2X placebo are presented in Figures 24–25, respectively. No interfering peaks from the diluent, 1X, and 2X placebo were observed.

TABLE 23   Forced Degradation (Degradants ≥ 0.05% w/w) (RT$_{compound}$ A = 6.11 min in the standard)

Forced Degradation (Degradants ≥ 0.05% w/w) (RT$_{compound}$ A = 6.11 min in the standard)	
Treatment	RRT
Spiked placebo acid degradation with heat (24 hrs)	0.22, 0.18, 0.19, 0.23, 0.31, 0.33, 0.35, 0.36, 0.39, 0.44, 0.46, 0.63, 0.74, 0.86, 0.91, 0.96, 1.13, 1.22, 1.32, 1.38, 1.51, 1.63, 1.70, 1.71, 1.87, 1.94, 1.97, 2.09
Spiked placebo base degradation with heat (24 hrs)	0.22, 0.17, 0.19, 0.21, 0.24, 0.25, 0.29, 0.33, 0.35, 0.52, 0.96, 1.75, 1.87
Spiked placebo peroxide with UV irradiation (24 hrs)	0.39, 0.41, 0.49, 0.53, 0.63, 0.96, 1.17, 1.41
Spiked placebo pyrolysis in aqueous solution (22 hrs)	0.22, 0.19, 0.22, 0.26, 0.28, 0.33, 0.37, 0.43, 0.49, 0.55, 0.60, 0.68, 0.70, 0.75, 0.97, 0.81, 0.84, 0.88, 0.89, 0.93, 0.96, 1.07, 1.09, 1.13, 1.17, 1.24, 1.27, 1.32, 1.45, 1.50, 1.53, 1.58, 1.64, 1.73, 1.80, 1.87, 1.93, 1.96, 1.89
Spiked placebo UV irradiation (48 hrs)	None

Reference: LAB BOOK 11 OWNER pp. 41, 43–54.

ii. Related Impurities/Degradants and Relative Response Factor Determination

1. A sample of each individual impurity was prepared and injected in duplicate. None of the impurities was found to elute in the elution zone of the active. Chromatograms are shown in Figures 26–27.
2. The relative response factors for related A and B were determined using the LOD/LOQ data. The slope of compound A obtained for the linear curves was divided by the slope obtained for related A and B in the LOD/LOQ experiments. Results are shown in Table 21.

g. Forced Degradation Study

i. The proper treatment times and conditions determined during the development work (report DR1234) are summarized in this section.
ii. The 5-mg placebo with FD&C Red #40 dye was used for the forced degradation study (worse case). A stock solution of compound A was also prepared at 5.0 mg/mL.

TABLE 24   Robustness—HPLC Parameter Variations (Compound A Peak)

Robustness—HPLC Parameter Variations (Compound A Peak)

Parameter Change	RT (min)	Theoretical Plate (N)	Tailing Factor (T)	Capacity Factor (k')
Method conditions	6.99	29505	1.1	11.4
Flow 1.35 mL/min (-10%)	6.69	28549	1.1	11.1
Flow 1.56 mL/min ($+10\%$)	6.43	26419	1.1	12.3
25 °C column temperature (-5 °C)	6.48	26990	1.1	11.6
35 °C column temperature ($+5$ °C)	6.77	29070	1.1	13.4
Mobile phase pH of 4.75 (-0.25 pH unit)	4.14	11474	1.1	7.5
Mobile phase pH of 5.25 ($+0.25$ pH unit)	7.36	22557	1.0	10.7
Mobile phase A (-10% buffer volume)	6.39	9976	1.1	6.8
Mobile phase A ($+10\%$ buffer volume)	12.74	67105	1.0	20.1
Mobile phase A (-10% buffer concentration)	10.04	25560	1.0	12.8
Mobile phase A ($+10\%$ buffer concentration)	9.80	24764	1.0	12.9
Mobile phase A (-10% acetonitrile volume)	10.39	32693	1.1	12.9
Mobile phase A ($+10\%$ acetonitrile volume)	8.25	18705	1.0	10.8
Mobile phase B (-10% THF volume)	8.97	28330	1.1	11.5
Mobile phase B ($+10\%$ THF volume)	8.44	27069	1.1	11.6

Reference: LAB BOOK 20 OWNER pp. 21, 41, 48, 55, 63, 76, 83, 90, 96, 102, 108, 114, 120, 130, 136, 143, 154, 221, 228, 175, 182, 189, 196, 203.
Specification: Report results

iii. Control (API)

1. A 5.0-mL aliquot of the compound A spiking solution was pipetted into a 50-mL volumetric flask and the solution was then diluted to volume with the method diluent (50:50 mobile phase A: THF). The final sample concentration was 0.1 mg/mL.

TABLE 25   Robustness—HPLC Parameter Variations (Impurities – Retention Time (RT) in minutes)

Robustness—HPLC Parameter Variations (Impurities – Retention Time (RT) in minutes)

Parameter Change	Related B		Related A	
	RT	RRT	RT	RRT
Method conditions	12.73	1.39	8.30	1.52
Flow 1.35 mL/min (–10%)	9.45	1.38	8.95	1.49
Flow 1.56 mL/min (+10%)	10.92	1.40	7.51	1.54
25 °C column temperature (-5 °C)	10.24	1.38	7.83	1.51
35 °C column temperature (+5 °C)	12.40	1.39	7.90	1.52
Mobile phase pH of 4.75 (−0.25 pH unit)	10.23	1.57	2.92	1.83
Mobile phase pH of 5.25 (+0.25 pH unit)	10.96	1.49	5.68	1.68
Mobile phase A (−10% buffer volume)	8.37	1.63	2.03	1.88
Mobile phase A (+10% buffer volume)	17.00	2.83	2.89	1.28
Mobile phase A (−10% buffer concentration)	12.63	1.38	8.13	1.51
Mobile phase A (+10% buffer concentration)	12.46	1.39	7.89	1.52
Mobile phase A (−10% acetonitrile volume)	9.91	1.43	9.13	1.43
Mobile phase A (+10% acetonitrile volume)	9.99	1.46	2.84	1.64
Mobile phase B (−10% THF volume)	12.46	1.37	7.99	1.50
Mobile phase B (+10% THF volume)	9.35	1.39	8.85	1.51

Reference: LAB BOOK 20 OWNER p. 205.
Specification: Report results.

TABLE 26   Robustness—HPLC Column Variation (Compound A)

Robustness—HPLC Column Variation (Compound A)

Column Serial Number	Retention Time (min)	Theoretical Plate Count (N)	Tailing Factor (T)	Capacity Factor (k')
M32122008 (new column)	6.99	29505	1.1	11.4
M22901F	7.02	27572	1.1	11.5
M30631S	6.80	23893	1.0	8.2

Reference: LAB BOOK 20 OWNER pp. 21, 28, 35.
Specification: The retention times should be similar on each column.

TABLE 27   Robustness—HPLC Column Variation (Impurities – Retention Time (RT) in minutes)

Robustness—HPLC Column Variation (Impurities – Retention Time (RT) in minutes)

Parameter Change	Related B		Related A	
	RT	RRT	RT	RRT
M32122008 (new column)	12.73	1.39	8.30	1.52
M22901F	12.70	1.39	8.26	1.52
M30631S	12.17	1.40	7.70	1.53

Reference: LAB BOOK 20 OWNER p. 206.
Specification: The retention times should be similar on each column.

   iv. Acid Degradation with Heat (API)
      1. A 5.0-mL aliquot of the compound A spiking solution was pipetted into a 50-mL volumetric flask and 5.0 mL of 0.5 N HCl was added. The solution was placed in an oven at 75 °C for 24 hours and then neutralized with 5.0 mL of 0.5 N NaOH. The solution was then diluted to volume with the method diluent (50:50 mobile phase A: THF).
   v. Base Degradation with Heat (API)
      1. A 1.0-mL aliquot of the compound A stock solution was pipetted into a 25-mL volumetric flask and 5.0 mL of 0.5 N NaOH was added. The solution was placed in an oven at 75 °C for 24 hours and then neutralized with 5.0 mL of 0.5 N HCl. The solution was then diluted to volume with the method diluent (50:50 mobile phase A: THF).

TABLE 28   Stability of an Assay Working Standard Solution (0.2 mg/mL)

Stability of an Assay Working Standard Solution (0.2 mg/mL)

Storage Condition	% Recovery	% Difference from Initial (Absolute)
Initial	100.0	NA
24-hour benchtop—clear	100.1	0.1
24-hour benchtop—amber	100.0	0.0
24-hour refrigerated—clear	99.9	0.1
24-hour refrigerated—amber	99.9	0.1
48-hour benchtop—clear	101.0	1.0
48-hour benchtop—amber	101.0	1.0
48-hour refrigerated—clear	100.9	0.9
48-hour refrigerated—amber	100.8	0.8

Reference: LAB BOOK 14 OWNER p. 36.
Specification: The recovery value does not vary more than 1.5% (absolute) from the initial result.

TABLE 29   Stability of a Sample Solution (~0.2 mg/mL)

Stability of a Sample Solution (~0.2 mg/mL)

Storage Condition	% Label Claim	% Difference from Initial (Absolute)
Initial	99.0	NA
24-hour benchtop—clear	89.8	0.2
24-hour benchtop—amber	89.9	0.1
24-hour refrigerated—clear	89.9	0.2
24-hour refrigerated—amber	99.0	0.1
48-hour benchtop—clear	99.6	0.5
48-hour benchtop—amber	99.7	0.7
48-hour refrigerated—clear	99.7	0.7
48-hour refrigerated—amber	99.9	0.9

Reference: LAB BOOK 14 OWNER p. 37.
Specification: The recovery value does not vary more than 1.5% (absolute) from the initial result.

TABLE 30 **Stability of an Impurity Working Standard Solution (0.0002 mg/mL or 0.1%)**

Stability of an Impurity Working Standard Solution(0.0002 mg/mL or 0.1%)			
Storage Condition	% Recovery	% Level from Nominal Concentration	% Difference from Initial % Level (Absolute)
Initial	100.0	0.10	NA
24-hour benchtop—clear	97.6	0.10	0.00
24-hour benchtop—amber	94.3	0.09	0.01
24-hour refrigerated—clear	94.3	0.09	0.01
24-hour refrigerated—amber	94.2	0.09	0.01
48-hour benchtop—clear	96.5	0.10	0.00
48-hour benchtop—amber	99.8	0.10	0.00
48-hour refrigerated—clear	90.8	0.09	0.01
48-hour refrigerated—amber	97.4	0.10	0.00

Reference: LAB BOOK 14 OWNER p. 38.
Specification: The level does not vary more than 0.25% (absolute) from the initial result.

TABLE 31 **Extraction Efficiency — Assay (5-mg Aged Tablets)**

Extraction Efficiency – Assay (5-mg Aged Tablets)			
Sonication Time (Minutes)	Preparation Number	% Label Claim	Average % Label Claim
10	1	97.3	97.3
	2	97.6	
	3	97.2	
15	4	97.1	97.1
	5	97.3	
	6	96.9	
30	7	97.6	97.2
	8	96.9	
	9	97.3	
5	10	97.4	97.4
	11	97.1	
	12	97.7	

Reference: LAB BOOK 12 OWNER p. 3.
Specification: Report results.

TABLE 32   Extraction Efficiency – Assay (10-mg Aged Tablets)

Extraction Efficiency – Assay (10-mg Aged Tablets)

Sonication Time (Minutes)	Preparation Number	% Label Claim	Average % Label Claim
10	1	96.2	96.2
	2	96.3	
	3	96.1	
15	4	95.7	95.7
	5	95.8	
	6	95.7	
30	7	95.7	95.7
	8	95.9	
	9	95.5	
5	10	96.5	96.3
	11	96.3	
	12	96.2	

Reference: LAB BOOK 12 OWNER p. 4.
Specification: Report results.

TABLE 33   Extraction Efficiency — Content Uniformity (5-mg Aged Tablets)

Extraction Efficiency – Content Uniformity (5-mg Aged Tablets)

Sonication Time (Minutes)	Preparation Number	% Label Claim	Average % Label Claim
10	1	95.0	95.2
	2	97.2	
	3	93.6	
15	4	96.3	95.6
	5	96.2	
	6	94.1	
30	7	94.7	95.2
	8	95.4	
	9	95.4	
5	10	94.9	95.5
	11	89.3	
	12	93.3	

Reference: LAB BOOK 12 OWNER pp. 7–10.

TABLE 34   Extraction Efficiency—Content Uniformity (10-mg Aged Tablets)

Extraction Efficiency – Content Uniformity (10-mg Aged Tablets)

Sonication Time (Minutes)	Preparation Number	% Label Claim	Average % Label Claim
12	1	94.2	
	2	96.7	95.7
	3	96.2	
20	4	94.4	
	5	94.5	94.7
	6	95.2	
30	7	95.1	
	8	96.1	96.2
	9	97.4	
5	10	95.4	
	11	92.9	94.0
	12	93.8	

Reference: LAB BOOK 12 OWNER pp. 11–14.

TABLE 35   LOD/LOQ Compound A

LOD/LOQ Compound A

Level	Avg S/N Ratio ($n = 5$)	Avg Area ($\mu V \bullet s$) ($n = 5$)	% RSD ($n = 5$)
0.005% (0.0104 µg/mL)	2	226.45	64
0.01% = LOD (0.0208 µg/mL)	4	522.35	57
0.02% (0.0415 µg/mL)	8	1304.30	13
0.05% = LOQ (0.1038 µg/mL)	21	3020.35	8

Reference: LAB BOOK 22 OWNER pp. 8, 10, 12, 14, 42, 61A.
Specification: The LOD is the first level producing a signal-to-noise ratio of 3:1. The LOQ is the first concentration at which the % RSD is 10% or less. The % RSD, for all concentrations greater than LOQ, should also be 10% or less. The LOQ is the level at which a signal-to-noise ratio of 10:1 is obtained, and the % RSD is less than 10%.

TABLE 36   LOD/LOQ Related A

	LOD/LOQ Related A		
Level	Avg S/N Ratio ($n = 5$)	Avg Area ($\mu V \bullet s$) ($n = 5$)	% RSD ($n = 5$)
0.005% (0.0106 µg/mL)	2	248.89	25
0.01% = LOD (0.0213 µg/mL)	3	488.60	8
0.02% (0.0426 µg/mL)	8	1067.45	4
0.05% = LOQ (0.1064 µg/mL)	20	2670.10	2

Reference: LAB BOOK 22 OWNER pp. 8, 10, 12, 13, 42, 61A.
Specification: The LOD is the first level producing a signal-to-noise ratio of 3:1. The LOQ is the first concentration at which the % RSD is 10% or less. The % RSD, for all concentrations greater than LOQ, should also be 10% or less. The LOQ is the level at which a signal-to-noise ratio of 10:1 is obtained, and the % RSD is less than 10%.

TABLE 37   LOD/LOQ Related B

	LOD/LOQ Related B		
Level	Avg S/N Ratio ($n = 5$)	Avg Area ($\mu V \bullet s$) ($n = 5$)	% RSD ($n = 5$)
0.005% = LOD (0.0104 µg/mL)	3	439.20	17
0.01% (0.0207 µg/mL)	5	681.80	8
0.02% = LOQ (0.0415 µg/mL)	10	1302.50	2
0.05% (0.1037 µg/mL)	25	3263.95	3

Reference: LAB BOOK 22 OWNER pp. 8, 10, 12, 14, 42, 61A.
Specification: The LOD is the first level producing a signal-to-noise ratio of 3:1. The LOQ is the first concentration at which the % RSD is 10% or less. The % RSD, for all concentrations greater than LOQ, should also be 10% or less. The LOQ is the level at which a signal-to-noise ratio of 10:1 is obtained, and the % RSD is less than 10%.

TABLE 38

System Suitability				
Retention Time (min) (compound A)	Theoretical Plate Count (N)	Tailing Factor (T)	Capacity Factor (k')	% RSD Assay Standard ($n = 5$)
6.80	29717	1.1	14.2	0.1

Reference: LAB BOOK 15 OWNER pp. 24, 24A, 24B.

Although no requirements were initially set in the proposed method, impurity standards at the 0.1% level should meet the following criteria for precision and standards agreement: % RSD for six replicate injections should be NMT 10% and the main/check agreement should be NMT 10%. The impurity standards were determined to meet these criteria throughout the validation. Example: As determined during the accuracy study, the precision for the impurity standard was 5.5% and the main/check agreement was 4.0% (LAB BOOK OWNER19 p. 18).

vi. **Hydrogen Peroxide Degradation with UV Irradiation (API)**

1. A 1.0-mL aliquot of the compound A stock solution was pipetted into a 25-mL volumetric flask and 5.0 mL of 3.0% hydrogen peroxide was added. The solution was exposed to 8 W/m^2 of UV irradiation for 48 hours. The solution was then diluted to volume with the method diluent (50:50 mobile phase A: THF).

vii. **UV Irradiation in Aqueous Solution (API)**

1. A 1.0-mL aliquot of the compound A stock solution was pipetted into a 15-mL quartz tube and 5.0 mL of deionized water was added. The solution was exposed to 8.5 W/m^2 of UV irradiation for 48 hours. The sample was then transferred to a 25-mL volumetric flask. The quartz tube was rinsed repeatedly with the method diluent (50:50 mobile phase A: THF) and the rinse solutions were combined with the sample solution in the volumetric flask. The solution was then diluted to volume with the method diluent (50:50 mobile phase A: THF).

viii. **Pyrolysis in Aqueous Solution (API)**

1. A 1.0-mL aliquot of the compound A stock solution was pipetted into a 25-mL volumetric flask and 5.0 mL of deionized water was added. The solution was placed in an oven at 95 °C for 24 hours. The solution was then diluted to volume with the method diluent (50:50 mobile phase A: THF).

ix. **Control (Placebo)**

1. A 150.0 mg sample 5 mg placebo was added to a 50-mL volumetric flask and the solution was then diluted to volume with the method diluent (50:50 mobile phase A: THF).

x. Acid Degradation with Heat (Spiked Placebo)

  1. A 1.0-mL aliquot of the compound A stock solution was pipetted into a 25-mL volumetric flask and 5.0 mL of 0.5 N HCl was added. Placebo (750 mg) was then transferred into the flask. The solution was placed in an oven at 75 °C for 24 hours and then neutralized with 5.0 mL of 0.5 N NaOH. Approximately 10 mL of the method diluent (50:50 mobile phase A: THF) was then added and the solution was sonicated for ~10 minutes. The solution was then diluted to volume with the method diluent (50:50 mobile phase A: THF). The solution was transferred to a 50-mL polypropylene centrifuge tube and centrifuged at 5000 rpm for ~10 minutes to obtain a clear solution.

xi. Base Degradation with Heat (Spiked Placebo)

  1. A 1.0-mL aliquot of the compound A stock solution was pipetted into a 25-mL volumetric flask and 5.0 mL of 0.5 N NaOH was added. Placebo (750 mg) was then transferred into the flask. The solution was placed in an oven at 75 °C for 24 hours and then neutralized with 5.0 mL of 0.5 N HCl. Approximately 10 mL of the method diluent (50:50 mobile phase A: THF) was then added and the solution was sonicated for ~10 minutes. The solution was then diluted to volume with the method diluent (50:50 mobile phase A: THF). The solution was transferred to a 50-mL polypropylene centrifuge tube and centrifuged at 3700 rpm for ~10 minutes to obtain a clear solution.

xii. Hydrogen Peroxide Degradation with UV Irradiation (Spiked Placebo)

  1. Placebo (750 mg) was transferred into a 15-mL quartz tube and a 1.0-mL aliquot of the compound A stock solution and 5.0 mL of 3.0% hydrogen peroxide was added. The solution was exposed to 8 W/m^2 of UV irradiation for 24 hours. The sample was then transferred to a 25-mL volumetric flask. The quartz tube was rinsed repeatedly with the method diluent (50:50 mobile phase A: THF) and the rinse solutions were combined with the sample solution in the volumetric flask. The solution was then diluted to volume with the method diluent (50:50 mobile phase A: THF). The solution was transferred to a 50-mL polypropylene centrifuge tube and centrifuged at 3700 rpm for ~10 minutes to obtain a clear solution.

xiii. UV Irradiation in Aqueous Solution (Spiked Placebo)

xiv. Placebo (750 mg) was transferred into a 15-mL quartz tube and a 1.0-mL aliquot of the compound A stock solution and 5.0 mL of deionized water were added. The solution was exposed to 8 W/m^2 of UV irradiation for 48 hours. The sample solution was then transferred to a 25-mL volumetric flask. The quartz tube was rinsed repeatedly with the method diluent (50:50 mobile phase A: THF) and the rinse solutions were combined with the sample solution in the volumetric flask. The solution was sonicated for ~ 10 minutes and then diluted to volume with the method diluent (50:50 mobile phase A: THF). The solution was transferred to a 50-mL polypropylene centrifuge tube and centrifuged at 3700 rpm for ~10 minutes to obtain a clear solution.

xv. Heat in Aqueous Solution (Spiked Placebo)

xvi. Placebo (750 mg) was transferred into a 25-mL volumetric flask and a 1.0-mL aliquot of the compound A stock solution and 5.0 mL of deionized water were added. The solution was placed in an oven at 105 °C for 22 hours. The solution was sonicated for ~25 minutes and diluted to volume with the method diluent (50:50 mobile phase A: THF). The solution was transferred to a 50-mL polypropylene centrifuge tube and centrifuged at 5000 rpm for ~10 minutes to obtain a clear solution.

Table 22 presents the percent degradation and the purity factor for the compound A peak, obtained under each condition. Peak purity analysis was performed with Hewey Dewey software. The compound A peak was found to be spectrally pure under each condition, i.e. the purity factor was greater than the purity threshold. The relative retention times (RRT) of all impurities at or above the 0.05% level (%w/w) are reported in Table 23 for the degraded spiked placebos. Chromatograms of the degraded spiked placebo overlaid with the degraded placebo under each condition are shown in Figures 28–37 along with their peak purity plots.

h. Robustness

i. Mobile Phase Variation/Temperature Variation/Flow Rate Variation/pH Variation

1. Method robustness was demonstrated by variation of the mobile phase composition, temperature, flow rate, and pH from the parameters specified in the method

2. Mobile phase composition varied by ± 5% and ± 10% of each component

3. Column temperature varied by ± 5 °C

4. Flow rate varied by ± 10% and ± 25%

5. pH varied by ± 0.25 pH units

A system suitability standard (0.2 mg/mL), spiked with each impurity at their limit, was prepared and injected in duplicate with each parameter change. This solution was used to perform robustness for both assay and impurity. The sytem suitability results are summarized in Table 24. The retention time (RT) and the relative retention time (RRT) for each known impurity are reported in Table 25. Examination of the results from Tables 24 and 25 suggest that the variation in buffer volume is a critical parameter since related B (with +10% buffer volume) did not elute within the method run time. Variations in buffer volume and pH also produced the largest variations in relative retention time.

ii. Column to Column Variation

Method robustness was demonstrated using three different analytical columns. A new column and two older columns were evaluated, all with different lots of packing material. The system suitability standard (0.2 mg/mL), spiked with all five impurities at their limit, was injected twice on each column and system suitability parameters were determined. Results are presented in Tables 26 and 27. For each column, the system suitability criteria as described in the method were met and retention times were similar on each column.

i. Solutions Stability Studies

The solution stability of an assay standard, an impurity standard, and a sample solution were evaluated at room temperature and refrigerated conditions in both clear and amber glassware. These samples were analyzed in duplicate against a fresh quantitation standard at initial, 24-hour, and 48-hour intervals. Results are presented in Tables 29, 30, and 31. The impurity standard was prepared per the proposed method at the 0.1% level. Therefore, the specifications for low-level impurity/degradants were applied to the impurity standard in Table 31. All solutions were determined to be stable for compound A for at least 48 hours under all storage conditions. The stability of impurities standard solution is shown in Table 30.

j. Extraction Efficiency

i. Assay

Twelve sample weights were prepared for the 5-mg and 10-mg aged samples. The 12 preparations were divided into four sets of three preparations. The extraction (sonication) time was varied for each set of three preparations and the percent label claims were calculated. The results are presented in Tables 31 and 32. Based on these results, the time range for sonication is 5–30 minutes for both the 5-mg and 10-mg strength. Thus, an acceptable extraction range can be chosen to be 5–10 minutes.

ii. Content Uniformity

Twelve tablets were prepared for the 5-mg and 10-mg aged samples per the content uniformity preparation. The 12 preparations were divided into four sets of three preparations. The extraction (sonication) time was varied for each set of three preparations and the percent label claims were calculated. The results are presented in Tables 33 and 34. Preparation # 1 for both the 5-mg and 10-mg strengths was also monitored to determine the disintegration times. The 5-mg tablet disintegrated after 5 minutes of sonication. The 5-mg tablet disintegrated after 10 minutes of sonication. Based on these results, the time range for sonication is 5–30 minutes for the 5-mg, and 5–30 minutes for the 10-mg strength. Thus, an acceptable extraction range can be chosen to be 10–15 minutes.

k. Limit of Detection (LOD) and Limit of Quantitation (LOQ)

The LOD and LOQ of the method were determined by evaluating solutions containing compound A and each impurity at several different concentrations (0.005%, 0.01%, 0.02%, and 0.05%). Five injections were made at each concentration and the % RSD and signal to noise ratios for compound A and for each impurity were determined. Tables 35–37 summarize the results obtained for compound A and B. Levels corresponding to LOD and LOQ are labeled for each table in the "Level" column.

l. System Suitability Determination

The system suitability parameters proposed in the draft method were met throughout the validation. An example of system suitability parameters, as determined during the accuracy study, is given in Table 38 for the main compound A peak.

V. CONCLUSION

An HPLC method has been validated for the assay, and impurities determined of compound A tablets, 5- and 10-mg. The method was shown to be selective, precise, linear, and accurate within the range of 50% to 150% of the nominal concentration for both tablet strengths. Nylon and PVDF filters were qualified for use with this method. The method was shown to be robust with the buffer volume and pH established as critical parameters. Standard (high and low level) and sample solutions were shown to be stable for at least 48 hours. All impurity solutions were determined to be stable for at least 48 hours under refrigerated conditions.

VI. ANALYTICAL METHOD

The following is a proposed method for the determination of the assay and impurities in compound A tablets, 5- and 10-mg, and compound A API

I. Reagents

Sodium acetate, HPLC grade
Glacial acetic acid, HPLC grade
THF, HPLC grade
Acetonitrile, HPLC grade
Water, in-house deionized water
Compound A reference standard

II. Mobile Phase Preparation

Mobile Phase A: 70:24:6 of 25 mM sodium acetate buffer (pH 5.0) Acetonitrile: THF
Mobile Phase B: 60:40 THF:acetonitrile
Preparation of 25 mM sodium acetate buffer, pH 5.0

Accurately weigh 9.6 g of sodium acetate and quantitatively transfer to a 5-L container. Add 5 L of purified water and mix until dissolved. Adjust the pH to 5.0 ($\pm$ 0.05) using glacial acetic acid. This solution preparation may be scaled up as necessary.

Preparation of mobile phase A (70:24:6::sodium acetate buffer (pH 5.0): acetonitrile:THF (v:v:v))

Prepare mobile phase A to have a ratio of 70:24:6 sodium acetate buffer (pH 5.0):acetonitrile:THF. For example, to prepare 4 L, combine 2800 mL of sodium acetate buffer (pH 5.0), 960 mL of acetonitrile, and 240 mL of THF. Mix well and degas by sonication prior to use.

Preparation of mobile phase B (60:40::THF:acetonitrile)

Prepare mobile phase B to have a ratio of 60:40 THF:acetonitrile. For example, to prepare 4 L, combine 2400 mL of THF and 1600 mL of acetonitrile. Mix well and degas by sonication prior to use.

Diluent (50:50 mobile phase A:THF)

Prepare a solution of mobile phase A and THF with a ratio of 50:50. For example, to prepare 4 L, combine 2000 mL of mobile phase A and 2000 mL of THF. Mix well.

III. Standard Preparation

<u>Assay standard solution (0.2 mg/mL) (prepare in duplicate)</u>

Accurately weigh ~20 mg of compound A reference standard into a 100-mL volumetric flask. Add ~50 mL of diluent and sonicate for 10–15 minutes with intermittent shaking until a clear solution is obtained. Allow solution to cool to room temperature, and dilute to volume with diluent. Label as main and check assay standards.

<u>Impurity standard solution (0.0002 mg/mL) (prepare in duplicate)</u>

Transfer 10.0 mL of the assay standard solution into a 1000-mL volumetric flask. Dilute to volume with diluent and mix well. Transfer 10.0 mL of this solution into a 100-mL volumetric flask. Dilute to volume with diluent and mix well. Label as main and check impurity standards.

<u>Sensitivity solution (0.0001 mg/mL)</u>

Transfer 10.0 mL of the assay standard solution into a 1000-mL volumetric flask. Dilute to volume with diluent and mix well. Transfer 5.0 mL of this solution into a 100-mL volumetric flask. Dilute to volume with diluent and mix well.

IV. Sample Preparation

<u>Assay and impurities sample preparation (prepare in duplicate)</u>

Accurately weigh 40 tablets, grind to fine powder, and calculate the average tablet weight. Accurately weigh duplicate samples of ~155 mg (of the powder equivalent to 5 mg or 10 mg of compound A) into separate 50-mL volumetric flasks. Fill approximately three by four full with diluent and sonicate for 10–15 minutes with intermittent shaking. Dilute to volume with diluent and centrifuge or filter a portion of the solution to obtain a clear solution.

<u>Drug substance (API) sample preparation</u>

Accurately weigh approximately 20 mg of API into a 100-mL volumetric flask. Add approximately 50 mL of diluent and sonicate for 10–15 minutes with intermittent shaking until a clear solution is obtained. Allow solution to cool to room temperature, and dilute to volume diluent.

V. Chromatographic Conditions

Column:	Serta, Sleeper C4, 3.5 mm, 4.6 × 150 mm
Column temp:	30 °C

Mobile phase A:	70:24:6::25 mM sodium acetate buffer (pH 5.0):acetonitrile:THF (v:v:v)
Mobile phase B:	60:40::THF:acetonitrile (v:v)
Flow rate:	1.50 mL/min
Detector:	UV at 254 nm
Injection volume:	50 μL
Run time:	35 min
Gradient profile:	(ramps are linear)

Time (min)	% A	% B
0	100	0
4	100	0
26	51	49
27	100	0
35	100	0

Assay

Assay standards and sample preparations may be analyzed according to the following injection sequence:

Test Solution	# of Injections
Blank (diluent)	2
Main standard	5
Check standard	2
Assay sample 1	1
Assay sample 2	1
Assay sample X	1
Main standard	1

Note: Inject samples, with no more than 10 sample injections between bracketing standard solutions.

Impurities

Impurity standards and sample preparations may be analyzed according to the following injection sequence:

Test Solution	# of Injections
Blank (diluent)	2

Sensitivity solution	1
Main impurity standard	6
Check impurity standard	2
Impurity sample 1	1
Impurity sample 2	1
Impurity sample X	1
Main impurity standard	1

Note: Inject samples, with no more than 10 sample injections between bracketing standard solutions.

Assay and Impurities

Standards and sample preparations may be analyzed according to the following injection sequence:

Test Solution	# of Injections
Blank (diluent)	2
Sensitivity solution	1
Check impurity standard	2
Main impurity standard	6
Check standard	2
Main standard	5
Main impurity standard	1
Sample 1	1
Sample 2	1
Sample X	1
Main standard	1
Main impurity standard	1

Note: Inject no more than 8 samples between bracketing standards.

VI. System Suitability

System suitability criteria will be established after completion of the validation. The following minimum criteria should be met for the assay standard.

- $k' \geq 5.0$, where k' is the capacity factor
- $T \leq 2$, where T is the tailing factor

- $R \geq 2.0$, where R is the resolution between adjacent peaks
- $N \geq 15000$, where N is the number of theoretical plates
- % RSD ≤ 2.0, where RSD is the relative standard deviation of five replicate standard injections
- Main check standard agreement $\pm 2.0\%$

Requirements for the impurity standards are:

- % RSD ≤ 10.0, where % RSD is the percent relative standard deviation of six replicate standard injections
- Main check standard agreement $\pm 10.0\%$, required only if run is for impurity testing only. Not needed if assay and impurity testing run together
- S/N 10 for the compound A peak in the sensitivity solution

VII. Calculations

Compound A assay for tablets:

$$\text{Compound A (mg/tablet)} = \frac{A_{\text{sam}}}{A_{\text{std}}} \times \frac{W t_{\text{std}} \times PF}{100 \text{ mL}} \times \frac{50 \text{ mL}}{W t_{\text{sam}}} \times W t_{\text{avg}}$$

$$\text{Compound A (\%label)} = \frac{\text{Compound A (mg/tablet)}}{TS} \times 100$$

Compound A assay for API:

$$\text{Compound A (\%w/w)} = \frac{A_{\text{sam}}}{A_{\text{std}}} \times \frac{W t_{\text{std}} \times PF}{100 \text{ mL}} \times \frac{100 \text{ mL}}{W t_{\text{sam}}} \times 100$$

% Impurity for tablets:

$$\%\text{Impurity (\%w/w)} = \frac{A_{\text{imp}}}{A_{\text{istd}}} \times \frac{W t_{\text{std}} \times PF}{W t_{\text{sam}}} \times \frac{W t_{\text{avg}}}{TS}$$

$$\times \frac{50 \text{ mL} \times 10 \text{ mL}}{100 \text{ mL} \times 1000 \text{ mL}} \times RRF \times 100$$

% Impurity for API:

$$\%\text{Impurity } (\%\text{w/w}) = \frac{A_{imp}}{A_{istd}} \times \frac{Wt_{std} \times PF}{Wt_{sam}}$$

$$\times \frac{100 \text{ mL} \times 10 \text{ mL}}{100 \text{ mL} \times 1000 \text{ mL}} \times RRF \times 100$$

where

A_{sam} = area response of compound A peak in the sample preparation

A_{imp} = area response of an impurity peak in the sample preparation

A_{std} = average area response of compound A peak in the standard preparation

A_{istd} = average area response of compound A peak in the impurity standard preparation

Wt_{std} = weight of standard, mg

Wt_{sam} = weight of powder taken in sample, mg

Wt_{avg} = average weight of tablets, mg

PF = standard purity factor

TS = tablet strength, mg

RRF = relative response factor (RRF = 1 for unknown impurities)

VII. PROTOCOL DEVIATIONS

PDR #0418, #0419 and #0422: For the filtering study, the protocol requires the samples to be centrifuged at 2000 rpm for 10 minutes. Under these conditions, it was not possible to obtain a clear solution and therefore the centrifugation conditions were changed to 5000 rpm for 30 minutes. Under these adjusted conditions, a clear solution was obtained.

PDR #0421: For the photolysis experiments, quartz cells were used as reaction containers rather than the stated scintillation vials. UV light is not transmitted through glass.

VIII. FIGURES

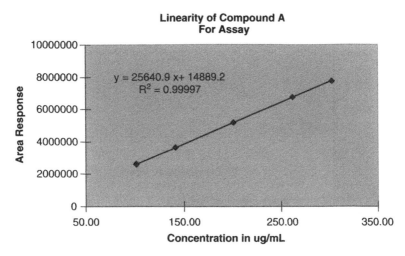

FIGURE 1 Linearity compound A for assay (50%–150% of nominal sample concentration).

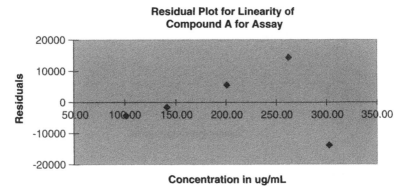

FIGURE 2 Residual plot for compound A for assay (50%–150% of nominal sample concentration).

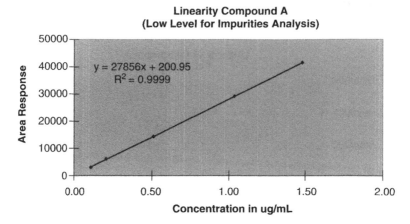

FIGURE 3 Linearity of response for compound A at low levels (0.05%–0.75% of nominal sample concentration).

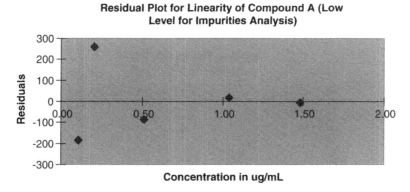

FIGURE 4 Residual plot for linearity of compound A at low levels (0.05%–0.75% of nominal sample concentration).

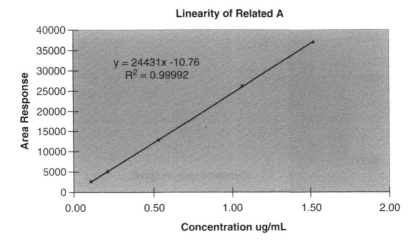

FIGURE 5   Linearity of related A.

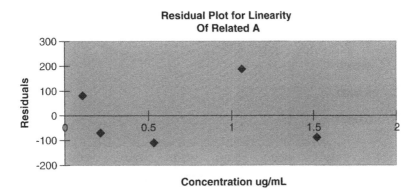

FIGURE 6   Residual plot for linearity of related A.

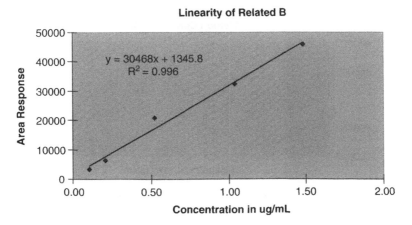

FIGURE 7   Linearity of related B.

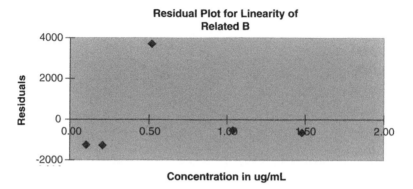

FIGURE 8   Residual plot for linearity of related B.

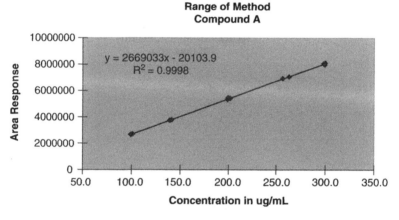

FIGURE 9   Range of method for compound A (LOQ – 150%).

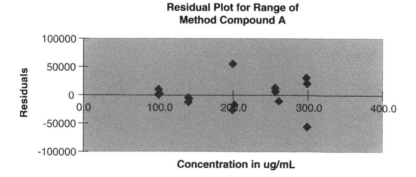

FIGURE 10   Residual plot from range of method for compound A (50% – 150%).

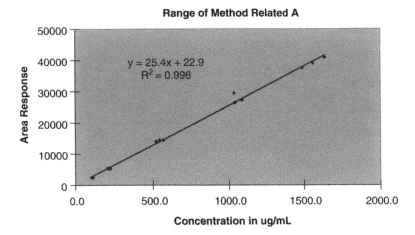

FIGURE 11 Linearity of method for related A (LOQ – 0.75%).

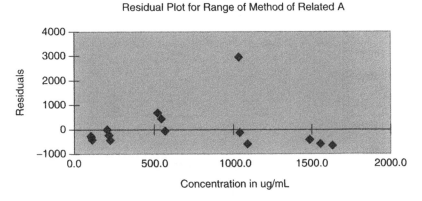

FIGURE 12 Residual plot from range of method for related A (LOQ – 0.75%).

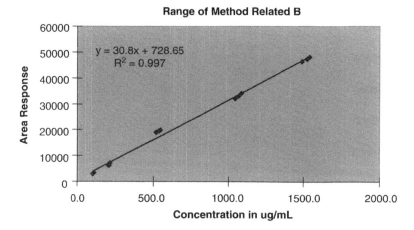

FIGURE 13 Range of method for related B (LOQ – 0.75%).

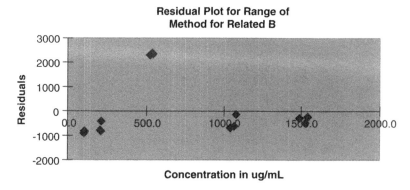

FIGURE 14 Residual plot from range of method for related B (LOQ – 0.75%).

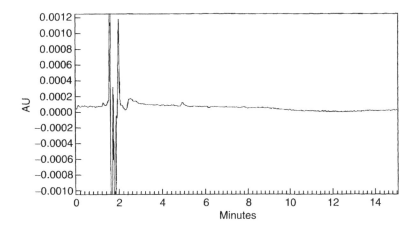

FIGURE 15   Typical blank chromatogram (diluent).

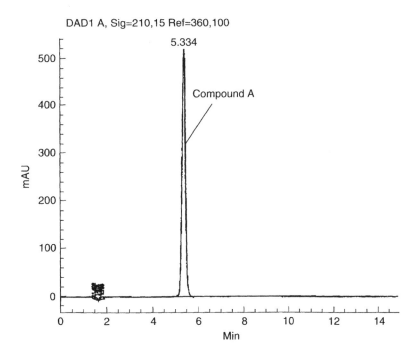

FIGURE 16   Chromatogram of an assay working standard (0.2 mg/mL).

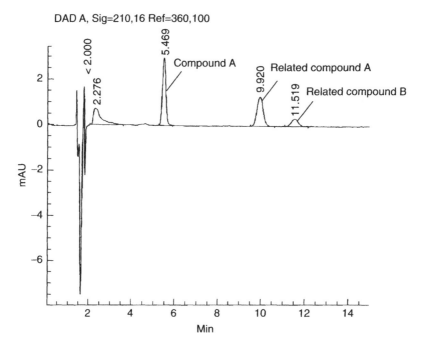

FIGURE 17 Chromatogram of an impurity standard (0.0002 mg/mL or 0.1%) spiked with related A and B.

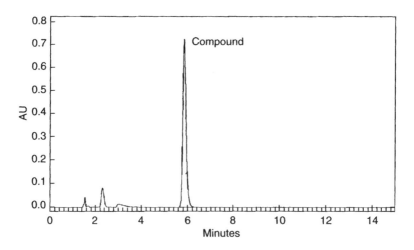

FIGURE 18 Chromatogram of a 5-mg sample preparation.

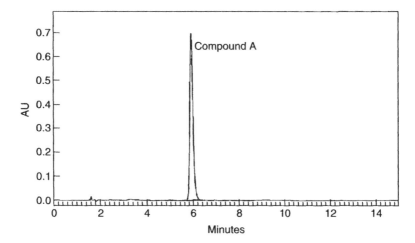

FIGURE 19   Chromatogram of a 10-mg sample preparation.

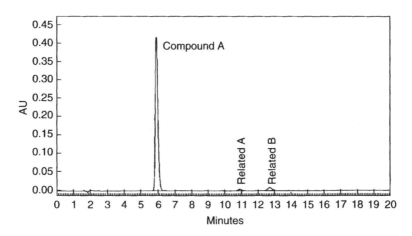

FIGURE 20   Chromatogram of a typical 5-mg sample preparation spiked with impurities at 0.1%.

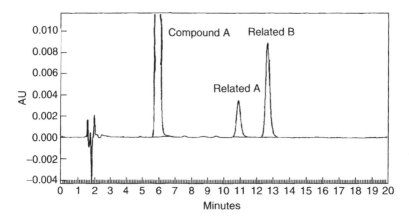

FIGURE 21 Expanded chromatogram of a 5-mg sample preparation spiked with impurities at 0.1%.

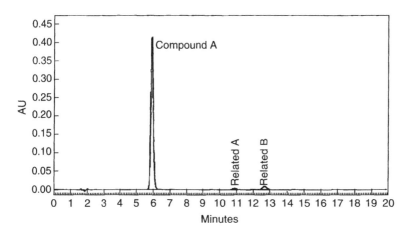

FIGURE 22 Chromatogram of a 10-mg sample preparation spiked with impurities at 0.1%.

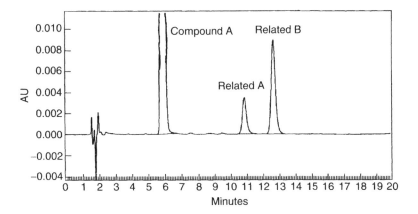

FIGURE 23 Expanded chromatogram of a 10-mg sample preparation spiked with impurities at 0.1%.

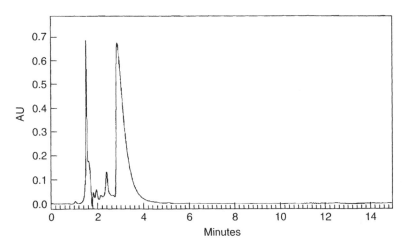

FIGURE 24 Chromatogram of 1× placebo + dye for the 5-mg strength (worst case).

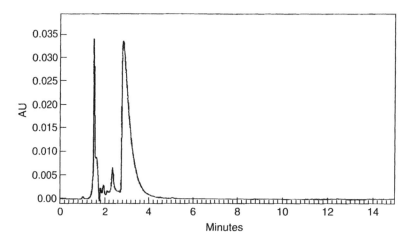

FIGURE 25 Chromatogram of 2 × placebo + dye for the 5-mg strength (worse case).

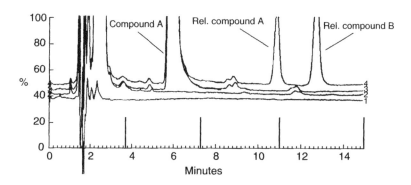

FIGURE 26 Comparison report representing diluent, 5.0-mg placebo, 5.0-mg placebo with active, and 5.0-mg placebo with active, related A and related B. Key: (1) diluent (bottom trace); (2) placebo; (3) placebo + compound A; and (4) placebo + compound A + related A + related B (top trace).

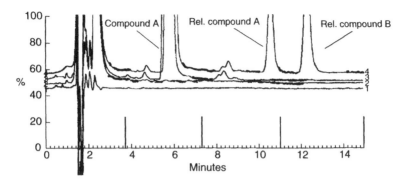

FIGURE 27 Comparison report representing diluent, 10.0-mg placebo, 10.0-mg placebo with active, and 10.0-mg placebo with active, related A and related B. Key : (1) diluent (bottom trace); (2) placebo; (3) placebo + compound A; and (4) placebo + compound A + related A + related B (top trace).

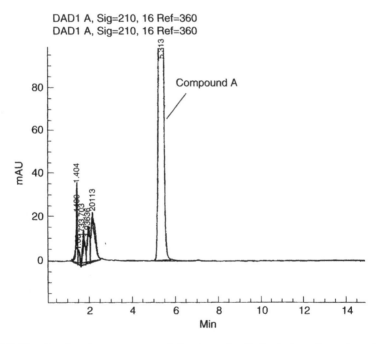

FIGURE 28 Overlay chromatograms of placebo and spiked solutions, force degraded with acid at 75 °C for 24 hours.

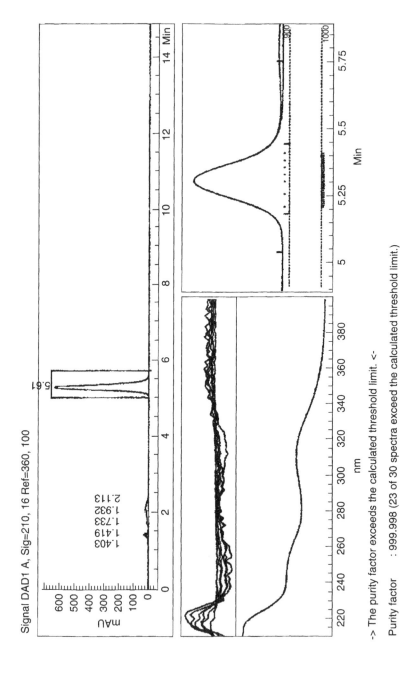

FIGURE 29 Spectral peak purity display for acid degraded spiked placebo.

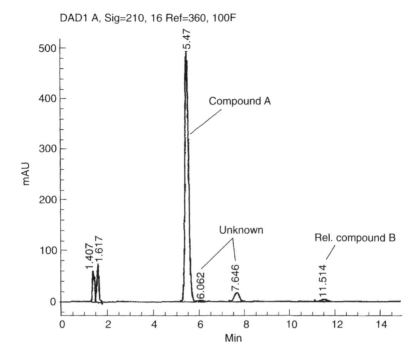

FIGURE 30 Overlay chromatograms of placebo and spiked placebo solutions, force degraded with base at 75 °C for 24 hours.

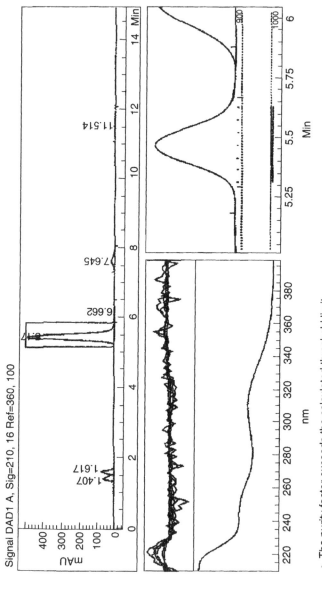

Signal DAD1 A, Sig=210, 16 Ref=360, 100

-> The purity factor exceeds the calculated threshold limit. <-

Purity factor   : 999.998 (23 of 47 spectra are within the calculated threshold limit.)
Threshold   : 999.99 (calculated with 23 of 47 spectra)
Reference   : Peak strat and end spectra (integrated) (5.299 / 5.672)
Spectra   : 7 (selection set by user, 7)
Noise threshold : 0.027 (12 spectra, St.Dev 0.0115 + 3* 0.0051)

FIGURE 31   Spectral peak purity display for base-degraded spiked placebo.

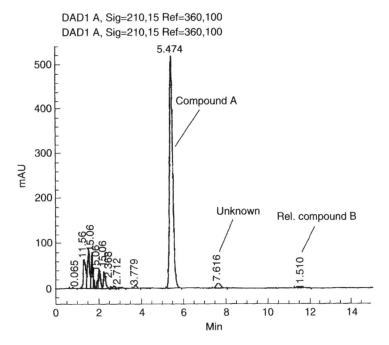

FIGURE 32   Overlay chromatograms of placebo and spiked placebo solutions, force degraded with hydrogen peroxide for 24 hours.

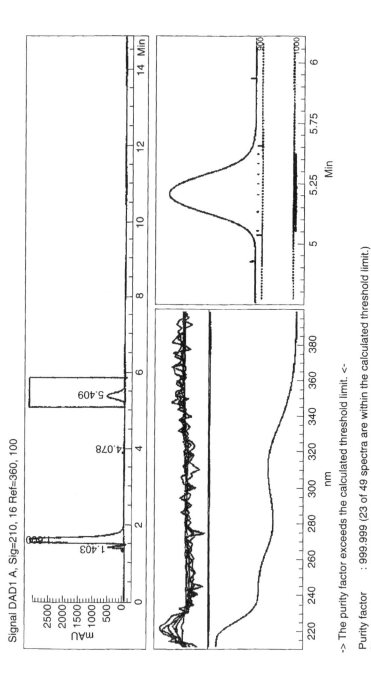

FIGURE 33 Spectral peak purity display for hydrogen peroxide-degraded spiked placebo.

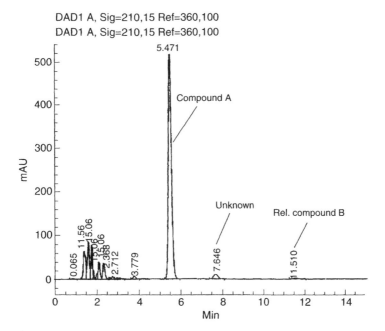

FIGURE 34   Overlay chromatograms of placebo and spiked placebo solutions, force degraded by UV exposure for 48 hours.

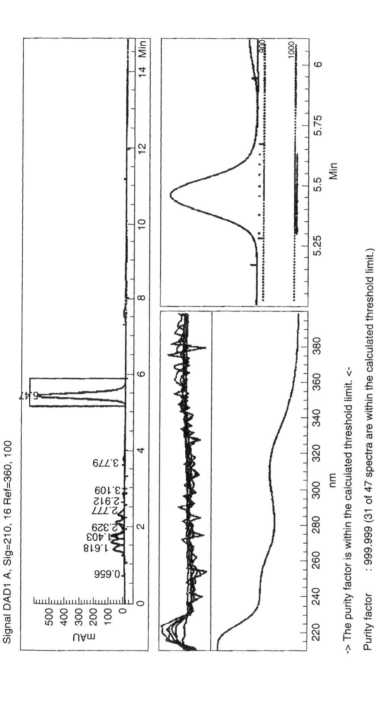

Signal DAD1 A, Sig=210, 16 Ref=360, 100

-> The purity factor is within the calculated threshold limit. <-

Purity factor   : 999.999 (31 of 47 spectra are within the calculated threshold limit.)
Threshold   : 999.999 (calculated with 31 of 47 spectra)
Reference   : Peak start and end spectra (integrated) (5.301 / 5.674)
Spectra   : 7 (selection set by user, 7)
Noise threshold   : 0.027 (12 spectra, St.Dev 0.115 + 3* 0.0051)

FIGURE 35 Spectral peak purity display for hydrogen peroxide-degraded spiked placebo.

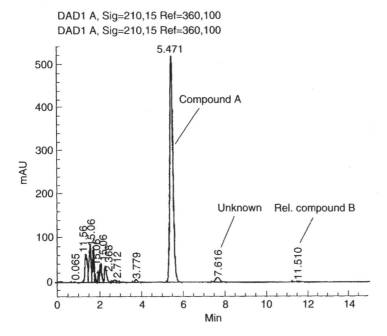

FIGURE 36   Overlay chromatograms of placebo and spiked placebo, force degraded by heat exposure (105 °C) for 22 hours.

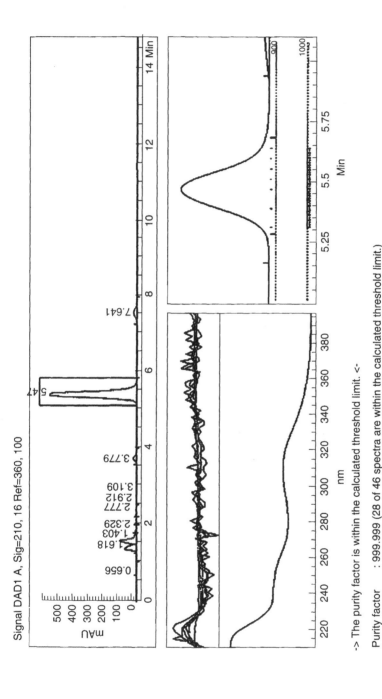

FIGURE 37 Spectral peak purity display for heat-degraded spiked placebo.

REFERENCES

1. Guidance for Industry: *Analytical Procedures and Methods Validation*, Draft, August 2000.
2. Chapter <1225> Validation of Compendial Procedures, *US Pharmacopoeia* (current), United States Pharmacopoeia Convention, Inc., Rockville, MD, 2006.
3. *Analytical Profiles of Drug Substances and Excipients*, Academic Press, New York, 2006.
4. *Code of Federal Regulations, Food and Drugs, Title 21 Part 211*, "Current Good Manufacturing Practices of Finished Pharmaceuticals," 2006.
5. ICH Q2A, *Text on Validation of Analytical Procedures*, March 1995.
6. ICH Q2B, *Validation of Analytical Procedures: Methodology*, May 1997.
7. Chapter <621> Chromatography, *US Pharmacopoeia* (current), United States Pharmacopoeia Convention, Inc., Rockville, MD, 2006.
8. AOAC *Book of Methods*, current.

INDEX

Acceptance criteria:
- characterized, 1–2, 46, 57
- sample methods validation report, 237
- sample protocol, 175–181, 183–188, 190, 192–193
- sample standard operating procedures (SOPs), 81, 95, 97–153

Accuracy:
- sample methods validation report, 228–230, 232–233
- sample preparation diagrams, 203–205
- sample standard operating procedures (SOPs), 81, 88, 93–99, 116–118, 138–139
- significance of, 9–11, 57

Acid degradation, 244
Action level, 57
Active pharmaceutical ingredient (API), 10, 58, 72, 81, 91–92, 240
Active substances, 90
Alert level, 57
Ambient conditions, 135, 212
Analysts, functions of, 10
Analytical methods validation, vii, 255–258
Analytical performance characteristics, 9, 58, 81, 88–92
API manufacturer, 6
Approval process, 1, 37, 170

Assay(s), *see also* Accuracy
- USP method category I, 96–153
- validations, 8
- working standard, 267

Atypical result, 58
Audit, 58, 62–63
Audit Summary Report (ASR), 58

Background, importance of, 8–9
Base degradation, 244
Baseline noise, 11
Batch, defined, 58
Batch record, 58
Biased method, 9
Blank chromatogram, 58, 81, 141, 267
Buffer/buffering agent, 58
Buffer pH variation, 107, 129–130, 147, 187

Calculations, sample protocol, 199–201
Calibration, 58
Calibration curve:
- implications of, 58–59, 81
- linearity, 91

Capacity, 12
Capacity factor k', 59, 81, 91, 105, 107, 127, 130, 145, 148
Carryover, 102, 142

Validating Chromatographic Methods. By David M. Bliesner
Copyright © 2006 John Wiley & Sons, Inc.

Prevalidation studies, 5
Prioritizing, 5
Process impurity, 65, 84
Product families, 134–135
Project plan:
 approval of, 37
 characteristics of, 24–26
 creation of, 45
 execution of, 38, 46
Prospective validation, 65
Protocol:
 defined, 65–66, 84
 deviations, 2
 example, 4
 sample, 169–217
 work plan, 39
Pyrolysis, 108, 131, 240, 250

Q., defined, 66, 84
Qualification, defined, 66
Quality assurance/Quality assurance unit (QA):
 defined, 66
 formal data review and report issuance, 53–54
 functions of, 45, 49–54, 75, 77
 importance of, 1–2, 5–6
Quality control (QC), 2, 5, 66
Quality system, defined, 66
Quantitation limit (QL):
 characteristics of, 9, 11
 defined, 66, 84
 sample protocol, 183–184
 sample standard operating procedures (SOPs), 90, 93–96, 102, 123, 142, 152
Quantitative tests, 96
Quarantine, 66
Questionnaire, end-user requirements, 154–155

R&D scientists, 6
Range, *see also* Linearity of method
 defined, 67, 84
 implications of, 9, 12
 sample methods validation report, 225, 229–231, 266
 sample protocol, 185–186
 sample standard operating procedures (SOPs), 90–91, 93–96, 103–105, 143–145, 152
Raw data, 67, 84, 194
Raw material, 67
Reagent(s):
 sample protocol, 195
 supplier, 6
Recovery/recoveries, *see also* Accuracy
 characteristics of, 88, 92, 97
 values, 9–10

Reference(s):
 sample, 79–80, 194
 standard, 67, 84
Regression line, 11. *See also* Linear regression analysis
Regulatory framework, vii, 5
Related compounds, 67, 84, 96
Relative response factor (RRF), 67, 85
Relative retention time (RRT):
 defined, 67
 sample methods validation report, 252–253
 sample standard operating procedures (SOPs), 85, 101–102, 121, 142, 152
Reliability, 91
Repeatability:
 defined, 65, 67, 83, 85
 sample methods validation report, 233–237
 sample preparation diagrams, 207–208
 sample protocol, 179
 sample standard operating procedures (SOPs), 88, 93, 95, 99, 118–119, 139–140, 151
 significance of, 10
Report(s), sample, 4, 194, 218–282
Reporting limit, 67, 85
Reprocessing, 67
Reproducibility, 65, 84, 107, 130, 148
Residuals/residuals plots:
 sample methods validation report, 261–265
 sample protocol, 176–177
 sample standard operating procedures (SOPs), 97–98, 102–104, 123, 125–126, 138–139,
Resolution:
 defined, 67–68, 85
 implications of, 12
 sample standard operating procedures (SOPs), 91, 105, 107–108, 127, 130, 145, 148
Resource allocation, 5
Response factor, 68, 85
Retain sample, 68
Retention time (RT):
 sample methods validation report, 242–243, 252
 sample standard operating procedures (SOPs), 91, 105, 127–128, 130, 145–146
 significance of, 12
Retrospective validation, 68
Revalidation, 68, 85
Review report/summary, 35
Robustness:
 defined 68, 85
 implications of, 9, 12, 25
 sample methods validation report, 242–244, 252–253
 sample standard operating procedures (SOPs), 91, 93–96, 127–130, 145–148, 152
 testing, 91, 127–130, 145–148, 152, 186–188

Printed and bound by CPI Group (UK) Ltd, Croydon, CR0 4YY

16/04/2025

14658352-0001